Dynamics of
TETHERED
SPACE
SYSTEMS

ADVANCES IN ENGINEERING

A SERIES OF REFERENCE BOOKS, MONOGRAPHS, AND TEXTBOOKS

Series Editor

Haym Benaroya

Department of Mechanical and Aerospace Engineering
Rutgers University

Published Titles:

Dynamics of Tethered Space Systems, *A. P. Alpatov, V. V. Beletsky, V. I. Dranovskii, V. S. Khoroshilov, A. V. Pirozhenko, H. Troger, and A. E. Zakrzhevskii*

Lunar Settlements, *Haym Benaroya*

Handbook of Space Engineering, Archaeology and Heritage, *Ann Darrin and Beth O'Leary*

Spatial Variation of Seismic Ground Motions: Modeling and Engineering Applications, *Aspasia Zerva*

Fundamentals of Rail Vehicle Dynamics: Guidance and Stability, *A. H. Wickens*

Advances in Nonlinear Dynamics in China: Theory and Applications, *Wenhu Huang*

Virtual Testing of Mechanical Systems: Theories and Techniques, *Ole Ivar Sivertsen*

Nonlinear Random Vibration: Analytical Techniques and Applications, *Cho W. S. To*

Handbook of Vehicle-Road Interaction, *David Cebon*

Nonlinear Dynamics of Compliant Offshore Structures, *Patrick Bar-Avi and Haym Benaroya*

Dynamics of
TETHERED
SPACE
SYSTEMS

A. P. Alpatov
V. V. Beletsky
V. I. Dranovskii
V. S. Khoroshilov
A. V. Pirozhenko
H. Troger
A. E. Zakrzhevskii

CRC Press
Taylor & Francis Group
Boca Raton London New York

CRC Press is an imprint of the
Taylor & Francis Group, an **informa** business

CRC Press
Taylor & Francis Group
6000 Broken Sound Parkway NW, Suite 300
Boca Raton, FL 33487-2742

First issued in paperback 2017

ISBN 13: 978-1-138-11793-8 (pbk)
ISBN 13: 978-1-4398-3685-9 (hbk)

Library of Congress Cataloging-in-Publication Data

Dynamics of tethered space systems / authors, A. P. Alpatov ... [et al.].
    p. cm. -- (Advances in engineering)
  Includes bibliographical references and index.
  ISBN 978-1-4398-3685-9 (hardcover : alk. paper)
  1. Tethered space vehicles--Dynamics. I. Alpatov, A. P. (Anatolii Petrovich) II. Title. III. Series.

TL798.T47D96 2010
629.43'4--dc22
                                                  2010008249

Visit the Taylor & Francis Web site at
http://www.taylorandfrancis.com

and the CRC Press Web site at
http://www.crcpress.com

Dedication

The authors and publisher dedicate this book to the late Dr. Hans Troger, who passed away in February 2010.

A native of Austria, Dr. Troger studied at the Vienna University of Technology, receiving his Master's degree there in 1966, and his Ph.D. in 1970. He went on to become an assistant professor, full professor, and Dean of Mechanical Engineering at V.U.T.

Dr. Troger was very interested in new developments in mathematics and their applications to mechanical problems. He applied the concepts of catastrophe theory and nonlinear stability theory to different fields in mechanics, such as the space tether dynamics, mode jumping of plates, buckling behaviour of shells, oscillations of fluid conveying tubes and railway bogies, motions of robot arms, just to name a few. In his book *Nonlinear Stability and Bifurcation Theory*, he tried to explain to engineers how they could apply these methods to their problems which he demonstrated with many examples from mechanical applications.

In order to learn the methods from their developers, Dr. Troger visited many international conferences and also motivated his co-workers to attend these conferences; he also organized conferences on "Bifurcation Theory" together with Tassilo Küpper, Friedemann W. Schneider, and Rüdiger Seydel. He invited researchers from abroad to Vienna, which resulted in close and productive collaborations with scientists from the East and West.

For his merits he was awarded an honorary doctorate by the Technical University of Budapest. The Austrian Academy of Science also awarded him the Schrödinger prize and elected him as member of the academy.

Dr. Troger will be remembered as a good friend and trusted colleague by all those who had the good fortune to work with him.

Contents

Preface

The idea to connect a system of bodies moving in space by long flexible cables goes back to K. E. Tsiolkovsky's work and hence has a history of more than a century. K. E. Tsiolkovsky also understood the important effect of small forces of gravity gradient on attitude motion and that artificial gravity could be created by rotation of a space system about its centre of mass.

Cable systems have been used since the beginning of space research in the second half of the 20th century. As examples we mention the "jo-jo type" of device, used on the American spacecraft *Transit-1A* in 1960, or the use of a cable by which the astronaut Leonov was connected to the spacecraft *Voskhod-2* in 1966. In 1966 during the flight of the manned spacecrafts *Gemini 11* and *Gemini 12* experiments were performed by connecting the spacecraft and the final stage of the launch vehicle *Agena*, for the purpose of studying the possibility of the formation of tethered systems. On *Gemini 11* the tether had been given a rotation with the angular velocity $0.0157\,\mathrm{rad\,s^{-1}}$, which was done during 1.5 circulations of the spacecraft in orbit around the Earth.

An experiment of a gravity-gradient stabilisation by means of a tether was again performed on the *Gemini 12* mission during three full circulations around the Earth. The crew expended about half of the time on deploying the tether in a gravitationally-stable state. Despite the fact that a control system of the spacecraft did not work, the problem of orientation of the tether was successfully solved.

The recent interest in tethered space systems (TSS) had its origin in the mid-seventies in connection with the proposed project of two Italian professors G. Colombo and M. D. Grossi (Smithsonian Astrophysical Laboratory, USA) to use a probe, hanging down on a 100 km long cable from a satellite in low Earth orbit, to collect data of the parameters of fields of the Earth and its atmosphere.

Cooperation between NASA on one side and the Italian Space Agency and the European Space Agency on the other side in preparing this project defined the main tasks for a successful development of the TSS concept. The new possibilities of the effective use of TSS achieved during the past ten years have made this concept one of the most promising new developments of astronautics. Novelty and originality of problems and research techniques of TSS behaviour attracted attention of experts all over the world. From the beginning of the eighties of the past century the subject of TSS started to form a separate area of space research.

At present the development of TSS is in the phase of studying full-scale experiments. After performing some test missions (TSS-1, TSS-1R, SEDS-1, SEDS-2, DMG, GHAGE, OEDIPUS) a certain level in the understanding of the basic features of the behaviour of TSS is reached. The experimental and theoretical research projects carried out within the last 25 years allowed to pass to real applications making use of the advantages of this concept. The achieved progress demonstrates great capabilities, on the one hand, for the feasibility of particular physical experiments and flight configurations, and on the other hand, for the proposal of new problems, whose consideration up to now seemed to be unjustified and premature.

Nowadays TSS belongs to a class of systems of various purposes, various design and great application potential. The dynamical problems encountered in the research of the motion of TSS and in the realisation of their design are even more diverse. The design process of a small independent TSS in joint work of the authors in the frame of the INTAS projects was the reason for the research documented in this book. Long-term and as it seems, fruitful cooperation of researchers not only from different geographical regions, but also from different fields of mechanics and space flight, resulted in a deeper understanding of the problems and selecting and working on the basic problems of the dynamics and development of TSS. An especially important and desirable aim is to connect successfully theoretical and experimental researches. However, in the book only material is included, which, according to the authors' common interest, relates to the field of mechanics. First of all, problems of non-linear dynamics and also the problems of development of physical and mathematical models of the dynamics of TSS are studied. In modern engineering, there exist whole classes of problems where the classical engineering approach, that is, building an experimental device and performing a series of tests, now is replaced by numerical simulation. Nevertheless there are processes which can be investigated experimentally much easier than by means of mathematical modelling and numerical simulation. Then one well-realised experiment gives much more information than dozens of theoretical investigations. Hence, some aspects of experimental investigations are included in this book. The selection of this material is determined by the general idea in writing this book, which is to present original subjects of common interest.

We hope that the book will succeed in giving the reader from the carried out analyses of the presented problems of TSS dynamics, the appropriate information for her or his own research, concerning both the methods of research and the combined application of theoretical and experimental methods.

About the Authors

A.P. Alpatov was born in 1941 in Barnaul, Russia, and studied at the Kazan Aviation Institute (Engineer 1965). He received his Ph.D. in Spacecraft Dynamics and Control (1972) from the Institute of Technical Mechanics NAS of the Ukraine, Dr. of Technical Science in Spacecraft Dynamics and Control (1988) from the Dniepropetrovsk State University. Professor (1992). Professional experience: 1966–1970 engineer in State Design Office "Youzhnoye"; 1970–present post-graduate, junior scientist, senior scientist, leading scientist, head of the department at the Institute of Technical Mechanics NAS&NSA of the Ukraine. Principal areas of research: spacecraft dynamics and control; mathematical modeling of complex systems.

V.V. Beletsky was born in 1930 in Irkutsk, Russia, and studied at Moscow State University, Department of Mechanics and Mathematics (Master 1954). He received a Ph.D. in Spacecraft Dynamics in 1962, a Dr. of Physics-Mathematics Science (1965) (both from the Keldysh Applied Mathematics Institute). Professional experience: 1954–present: research associate, senior research associate, head research associate at the Keldysh Applied Mathematics Institute, Moscow, Russia. 1962–present: Prof., Moscow State University, Department of Mechanics and Mathematics, part time position. Principal areas of research: space flight dynamics, celestial mechanics; robotics; nonlinear problems of mechanics including resonances and chaos. He is a member of the Russian Academy of Sciences (1997), a member of the International Astronautics Academy (1992), and was awarded the Alexander von Humboldt Prize (1992).

V.I. Dranovskii was born in 1934 in Dniepropetrovsk, Ukraine, and studied at the Dniepropetrovsk State University (DSU), (1957); he earned a Ph.D. in Spacecraft Dynamics (1969) and Dr. of Science (1992) both in DSU. Professional experience: 1957–1961: lecturer in DSU; 1961–2007: engineer, head of laboratory, head of department, main designer of spacecraft design Office in State Design Office "Youzhnoye". Principal areas of research: spacecraft dynamics and control; design of spacecraft. He is a member of the Ukrainian Academy of Sciences (2003).

V.S. Khoroshilov was born in 1939 in Russia and studied at the Kharkov Aviation Institute. (Engineer 1961); he earned his Ph.D. in 1968, and Dr. of Science in spacecraft dynamics and control, 1988 – both at the Dniepropetrovsk State University. Professional experience: 1961–present: engineer, senior scientist, head of sector in State Design Office "Youzhnoye". Principal areas of research: spacecraft dynamics and control.

A.V. Pirozhenko was born in 1961 in Krivoj Rog, Ukraine, and studied at the Dniepropetrovsk State University, (Master 1984); he received a Ph.D. in Theoretical Mechanics – the Institute of Mechanics NAS of Ukraine 1990; and a Dr. of Physics-Mathematics Science (2007) in Theoretical Mechanics at the Donetsk Institute of Applied Mathematics and Mechanics NASU. Professional experience: 1984–present: engineer, engineer-mathematician, leading mathematician, senior scientist, leading scientist at the Institute of Technical Mechanics NAS&NSA of the Ukraine. Principal areas of research: celestial mechanics; spacecraft dynamics and control, theoretical mechanics, theory of oscillations.

Hans Troger was born in 1943 in Austria, studied at the Vienna University of Technology (VUT) Mechanical Engineering (Master 1966 and PhD 1970), 1970 assistant professor and 1979 full professor for mechanics at VUT, 1985–87 he was Dean of Mechanical Engineering and a member of the Austrian Academy of Sciences (2002).

A.E. Zakrzhevskii was born in 1939 in Kiev, Ukraine, and studied at the Kiev Engineering-Building Institute (Engineer 1961); he earned a Ph.D. in Theoretical Mechanics from the Institute of Mechanics of Ukrainian Acad. of Sci, 1968; and Dr. of Technical Science in Spacecraft Dynamics and Control from the Moscow Aviation Institute (1985). Professional experience: 1961–64: engineer, junior scientist, senior scientist at the building design institutes, 1964–present: post-graduate, junior scientist, senior scientist, leading scientist, head of department at the S.P. Timoshenko Institute of Mechanics NAS of the Ukraine. Principal areas of research: spacecraft dynamics and control, theory of optimal control; theory of flexible multibody systems; dynamics of deployment of complex space systems.

Symbol Description

a	average value of distance r	$\vec{V}_{cm}$	initial velocity of mass centre
b	amplitude of longitudinal oscillations	$\vec{V}_1^{orig}$, $\vec{V}_{depl}$	velocity of separation of first body
c_i	arbitrary constants		
e	eccentricity of orbit	$\vec{V}_{dipl}^{fin}$	velocity of first body at instant of separation of second one
EF	tension stiffness of cable		
$\vec{e}_r$	unit vector directed along connecting line	$\vec{V}_{rot}$	velocity of separation of second body after deployment
$\vec{F}_i$	forces acting on end bodies	$\vec{V}_m$	velocity of separation of auxiliary mass
f	dimensionless tension of the cable		
i	inclination of orbit	$\vec{V}_1^{add}$	increment of velocity of first body
J	moment of inertia	X_1, Y_1, Z_1	
$l(t)$	current length of tether		absolute coordinates of mass centre of central body
l_1	initial length of tether		
l_2	nominal length of tether	β	angle between $\vec{V}_{cm}$ and the x axis
L	relative specific moment of momentum		
$\mathcal{L}$	normalized relative specific moment of momentum	μ	gravitational constant
		ν	true anomaly
m_i	masses of end bodies	ψ	precession angle
$\vec{m}^{C_1}$	moment of forces	θ	nutation angle
M	total mass	Θ^{C_1}	tensor of inertia of the central body with respect to its mass centre C_1
p	focal parameter of orbit		
$\vec{R}_k$	position vector of k-th body in absolute frame of reference		
$\vec{r}$	$= \vec{R}_2 - \vec{R}_1$	$\varphi_1, \varphi_2, \varphi_3$	Bryant's angles
$\vec{R}$	position vector of mass centre of tether about attractive centre	φ	pure rotation angle
		ω_0	angular velocity of orbital motion
T	tension force in the cable		
t_{depl}	duration of deployment	ω_i	angular velocity
t_p	duration of pushing apart of two bodies	ω_π	argument of the pericentre
		ω_{pass}^{max}	maximum achievable angular velocity
u	argument of latitude of orbital motion	Ω	longitude of ascending node

List of Figures

List of Tables

1

Tethered Systems in Space: A Short Introduction

1.1 Basic features and areas of applications

The concept of tethered satellite systems (TSS), that is, two or more satellites in orbit around a planet, connected by thin long cables — a length of 100 km is not unusual — is one of the most innovative concepts of satellite flight at the end of the 20th century. There exist numerous important practical applications [25], some of which were already tested in several flights in orbit around the Earth organized by NASA during the last decade of the 20th century. Some of these applications and, if applicable, the corresponding flights are shortly described in this report.

The large size of Tethered Systems in Space (TSS) extending from tens of meters up to hundreds of kilometers preserving mechanical, energetic and other connections between the end bodies is the basic difference of TSS to traditional Space systems. The use of cables to form extended Space systems results in a number of remarkable properties of TSS.

First, there is the possibility of interaction with the external fields of a planet. The moment of the gravitational forces acting on the tether depends on the square of the length of the connection. This allows to generate a highly stable radial relative equilibrium configuration of the deployed TSS on a circular orbit. In addition, artificial forces of gravity arise on the end bodies of the TSS resulting in a tension force in the tether. The gravy-gradient force on a mass, m, attached to the tether at a distance, Δr, from the system's centre of gravity is equal to the difference between the centrifugal and gravitational forces on it. An approximate value for this force [40] is given by, $F_{GG} \approx 3 \Delta r m \omega_o^2$. The value of the e.m.f. (electro-magnetic force), arising in a rectilinear conductor, due to the interaction with the magnetic field of the Earth is directly proportionally to the length of the conductor $E = -BlV_s$, where B is the magnetic induction, l is the length of the conductor, V_s is the velocity of the conductor, which is moving uniformly and perpendicularly to the force lines. Another purely mechanical property of TSS, is connected to the values of velocity V and centrifugal accelerations T_c

$$T_c = r\dot{y}^2, \ V = r\dot{y}, \ L = r^2\dot{y}, \tag{1.1}$$

1

where r is the distance between the bodies of the TSS, $\dot{y}$ is the angular velocity of rotation of the tether about the centre of masses, L is the specific moment of momentum of the tether. From (1.1) it follows, for example, that the subspacecraft deployed from the basic spacecraft downwards to the Earth will have the velocity of motion about the Earth by $\Delta R \omega_0$ smaller, than the velocity of motion of the mass centre, where ω_0 is the angular velocity of motion of the mass centre of the TSS in orbit. Hence the velocity of the subsatellite is essentially smaller than the velocity of a free spacecraft at this height. The square-law dependence of the centrifugal accelerations on the angular velocity allows to generate the slowly rotating TSS using the thin (and light) connecting string of length of hundreds of kilometers. The square-law dependence of the moment of momentum on the length of the connection can be used for the great ability of accumulation of moment of momentum and kinetic energy by the TSS. For example, a rotating TSS may be used for launching of a payload in an orbit as it is intended with the Skyhook or to capture and retrieve payloads as intended in the Tether Rendezvous System [40].

The creation of small TSS is also important. By small TSS we understand cable systems with a tether length from ten meters up to several kilometers with masses of attached bodies from one up to hundreds kg [7, 8]. The specific values of lengths and masses depend on the problem. The basic aspects of the concept "small TSS" consist, first, in the difference of the mass characteristics of these systems from the traditionally considered projects TSS-1 and TSS-2. Secondly, the concept "small TSS" means that its realisation is directed on the solution of some basic questions and thus the geometrical and mass characteristics and also its equipment are chosen such that possible expenses and risks are minimized.

The creation of small TSS is also of importance for itself. It can be used in plasma physics [101] when the distance between the collector and emitter in an experiment is 100–200 m. It is obvious that using a small rotating TSS for these purposes is more effective than the use, as it is proposed in [101], of a 100–200 meter long tower. The results of the Plasma Motor Generator (PMG) experiment serve as confirmation.

Interest for small TSS can be considerably increased by a strong increase of the role of microsatellites. The cable system "Spacecraft – microsatellite" comprises all profits of the use of cables for a Space station. The specific weight of the cable of about $1 \, \text{kg} \, \text{km}^{-1}$ corresponds fully to the requirements of small mass and dimensions of the microsatellites. The use of mechanical, informational and power connections will allow on the one hand to considerably reduce the equipment, and consequently the mass of the microsatellite, and will on the other hand considerably extend the class of problems solvable with the help of microsatellites.

Apparently, one of the basic modes of motion of small TSS with a length up to one kilometer is a fast (considerably faster than orbital) rotation of the system about the mass centre. As it was mentioned above, basic attention was given earlier to projects with gravitationally stabilised TSS. The Canadian project BICEPS – Bistatic Canadian Experiment on Plasma in Space [57] and the support by NASA [114] can be considered as a significant step in opening the direction of the use of rotating TSS. The opportunity of rather simple ways of variation of the length of the string, its tension and the angular velocity of rotation of the system have allowed to offer the use of rotating TSS for researches of Space plasma physics both of the high atmosphere and magnetosphere. Research of dynamics of such a system will allow to use the collected data for the project of creation of an artificial force of gravity.

Other opportunities of application of rotating TSS briefly are the following [44]. A rotating TSS used in the project of an aerodynamic probe would allow to lower a little the velocity of the motion of the probe in the atmosphere in comparison with the orbital rotation of the TSS. Thus, it is possible that the attached spacecraft making a rotatory motion about the mass centre will leave the dense layers of the atmosphere after immersing in them and a longer time will be in conditions, favorable for cooling. These conditions can be essential for the equipment, for example, carrying out supervision, photographing parts of the surface of the Earth. The rotating TSS passes in its rotation through different layers of the atmosphere and can scan them. Such research of the atmosphere can be an alternative to the research through a radially oriented STS having on the tether a set of "beads" of gauges [47, 114]. Use of rotating TSS expands significantly also the opportunity of their use for transport operations because both kinetic energy and moment of momentum of the TSS can be significantly increased.

Rotating the TSS about its mass centre can serve as an integrated gauge for research of the influences of fields of the Earth. TSS will allow to obtain an integrated, average estimation of the difference of influences on the attached bodies. On a half-revolution of rotation about the mass centre the influence will be accumulated as change of angular velocity of rotation. Hence the project of small TSS for research of latitudinal changes of density of the atmosphere, consisting in rotation a small TSS with two probe end bodies with different ballistic factors is interesting.

Small rotating TSS can serve as standard of length for calibrating and measurements of the characteristics of optical and radar-tracking systems located on board and on the Earth [128].

The question of using a rotating electrodynamical TSS has remained outside of the attention of researchers until today too. Although this question concerns more physics, it is obvious that use of rotating tethers in a magnetic field as Gertz's doublet will allow to generate in a separate conductor an alternating current. The positive solution of this question would open a perspective opportunity of the realisation of the project of an electrodynamical TSS in vacuum, i.e., without creation of a closed current in the ionosphere, without appropriate equipment and on higher orbits.

As it is visible, even a brief consideration of opportunities of rotating TSS shows that their use allows to achieve new effects practically in all areas of TSS use.

1.2   Physical models of TSS in literature

Choice of one or the other physical model of a TSS is determined by the structure of the TSS, the modes of its motion and the purposes of research. Hence the choice should be reasonable and the chosen physical model should correspond to the objectives of research.

Presently, developed mathematical models of TSS dynamics, despite their distinctions in details, may be divided into two classes:

- Class of models of TSS where the string is assumed to be a continuous visco-elastic medium connecting the satellites which are regarded as rigid bodies.

- Class of models where TSS is assumed to be a system of connected rigid bodies, i.e., the string is considered to be a massless mechanical connection and its own dynamics is not taken into account.

The model of a flexible string of linearly elastic material is usually used as a physical model for the cable. Taking into account bending and torsional stiffnesses of the string and its plastic deformation can have an essential effect on the dynamics of weakly loaded cables, and the modes of motion with large transverse oscillations. Hence, the axial line of a weakly loaded cable, just reeled off the coil, is a screw line — "a sucking-pig's tail" (Fig. 1.1)[38]. The stretching of the cable into a straight line often results in plastic deformations, the laws of which essentially differ from linear elasticity. Taking into account the stiffness of the cable for bending and twisting is necessary in investigations of formation of loops of the cable because these effects are important in such modes.

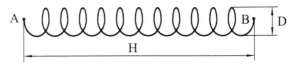

FIGURE 1.1
Sucking-pig's tail.

Modelling of the end bodies as a rigid body is not always acceptable, because, for example, the presence of extended elastic antennas at the end bodies in the OEDEPUS–A experiment could be one of the reasons for the rotational instability of the system. Usually the gravitational field of the spherical Earth is only used as a gravitational field, acting on the TSS. However, flatness of the Earth, attraction of Moon and Sun, and other effects can cause, for example, deviations of the TSS from the state of relative equilibrium. Hence, taking into account the flatness of a planet has allowed to consider the interesting project of perturbation of the orbital motion of the TSS from its relative equilibrium motion [34].

Aerodynamic resistance, light pressure, interaction of the STS with the magnetic field of the Earth and plasma, heating of the cable by solar radiation, collision with micrometeorites, all these effects can play an essential, and in some cases, determining role for the motion of the TSS. Their consideration in the model of motion of the TSS and the completeness of this model depends on the problem to be solved.

Selection of this or that physical model depends also on a mode of motion which makes TSS. The elementary classification of modes of motion, which is already found in the investigation of the motion of a mathematical pendulum is the separation of rotational and librational motions.

The librational motion of a tether of two bodies in the vicinity of the state of the stable radial equilibrium, as it was mentioned above, is the most investigated motion. For the assumption of small displacements and velocities of the motion the problem may be linearized and is reduced to the determination of stability of the equilibrium state in view of the various mentioned effects [130]. Hence, the research of the influence of the dynamics of the cable in such a formulation of the problem allows the use of methods of the most developed area of the mechanics of strings [1, 76, 100, 104], namely the theory of stationary motions. The problem is reduced to the analysis of the ordinary linear differential equations describing the relative equilibrium of the string instead of equations in partial derivatives describing general dynamics of the string.

The rotational motion of two bodies connected by a string at sufficiently large angular velocities of rotation of the system around the mass centre provides the essential tension in the string. Therefore, for this motion the transverse oscillations of the string will be small and their influence on the general motion of the system will be insignificant, and at least at the first step of investigation these fluctuations need not to be taken into account, i.e., the string for this mode of motion can be considered an elastic connection between the points of attachment of the end bodies. Since basic practical problems of investigation of rotational motions of STS are connected to the investigation of the change of parameters of rotation in view of various effects, the problem of investigation of these motions is reduced in essence to the investigation of the evolution of parameters of motion of distributed elastic systems.

If a significant redistribution of energy of motion is possible between its various mechanical forms, resonant modes of motion can serve on the one hand as the reason for destruction of the required configuration of motion of the system, and on the other hand can be used for both effective control and construction of stable motions of the system. In the motion of an orbital cable system of two bodies the resonant motion can arise at commensurability of the following average frequencies of motions: orbital motion of the mass centre, rotational (angular) motion of the TSS about its mass centre, translational relative (varying distance between bodies) motion, oscillations of the cable and rotational motions of end bodies about points of their attachment. It is obvious that the resonant commensurability of frequencies can take place for any pair of motions and for motions of the system in all degrees of freedom. Thus, the resonant motions represent a sufficiently wide and important area of possible motions of TSS. Up to the present time the area of resonant motions of TSS is only weakly investigated.

Another new phenomenon, which has been discussed intensively in the literature basically only over the last twenty years is the chaotic motion of deterministic systems. As it is well known (see for example [130]), this phenomenon is inherent to the majority of non-linear systems, and takes place already for systems with two frequencies of motions (enough one and half degrees of freedom). A classification of the given motions is necessary for the application of special numerical methods for the dynamics of TSS, since the usual numerical methods do not work in these cases and the usual mechanical interpretation of results of computations is not correct. This field of chaotic motions of TSS practically has not yet been investigated.

TSS are complex mechanical systems. Wide variety of modes of their motion has essential qualitative distinctions and is determined by specific ranges of values of their parameters. In fact, areas of librational and rotational motions, areas of regular and resonance trajectories, separatrices and chaotic motions possess essential qualitative features [21]. These features in many respects determine problems and methods of research and the range of values of parameters of TSS matching such regimes.

1.3   Comparison of the influence of various physical effects

We assume that the massive basic satellite B moves on a circular orbit at height $400\,\text{km}$ above the surface of the Earth and deployed from it, vertically towards the Earth, is a cable AB of length $l = 20\,\text{km}$ with the free end at A. The absence of an end mass allows to concentrate the attention only on the unperturbed forces F, acting on the cable. On the free end A the tension is equal to zero, and in point B it is maximum: $T_B \approx 1.5\rho\omega^2 l^2$, where ω is the angular velocity of rotation. (See [40], p. 123). In Fig. 1.2 levels of perturbations acting on the cable are shown: a denotes gravitational, b elastic, c electromagnetical and d other levels of the tension T_B depending on the diameter d of the cable made of corrosion-proof steel (continuous lines) and of kevlar (dot-and-dash lines).

The shaded strips correspond to a range of properties of heavier materials, for example of wolfram.

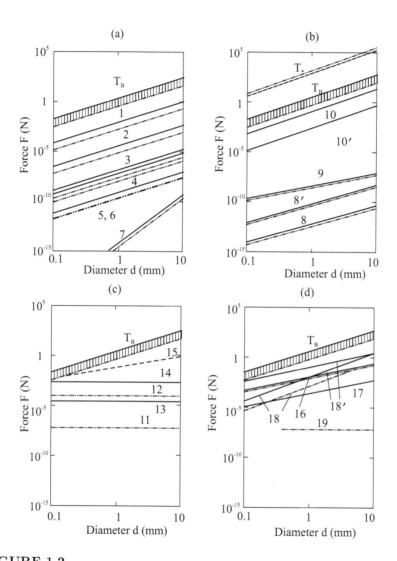

FIGURE 1.2

Magnitudes of perturbations, acting on the cable : *a* - gravitational, *b* - mechanical, *c* - electro-magnetic, *d* - others (the numbers in the figures correspond to the numbering of the various effects in the text.)

1.3.1   Gravitational perturbations.

1. Flattening of the Earth. A disturbing acceleration from the harmonic J_2 in standard representation of the geopotential [46] results in a change of the tension force of the cable δT only in connection with the difference of accelerations in different points of the TSS. If the disturbing acceleration in all TSS points were identical, the cable would not react on it by a change of its tension (situation "of free fall"). Integration along the cable of the gradient of the radial components of acceleration of J_2 gives $\delta T_B \approx 3\rho\omega^2 l^2 J_2 (R_\oplus/R_B)^2 (1-3\sin^2\varphi)$, where $R_\oplus$ denotes the radius of the Earth; φ is current geocentric latitude. The maximum increment of tension occurs in polar areas $\varphi = \pm\pi/2$. In equatorial areas and at latitude $\varphi = 0.96\,\text{rad}$ the relative increment of tension is $\delta T_B/T_B \approx 2J_2 \approx 2 \cdot 10^{-3}$ (line 1 in Fig. 1.2, full line — for a steel cable, dotted-and-dashed line — for a cable of kevlar).

2. Higher harmonics of the geopotential. They give a contribution to the tension not more than $\delta T_B/T_B \leq 10^{-5}$ (lines 2).

3. Attraction by the Moon. The maximum contribution of the gradient of the gravitational field of the Moon to the tension of the cable yields $\delta T_B \leq 2\rho G M_m R_m^{-3} l^2$ (G denotes the universal gravitational constant; M_m and R_m are mass and radius of the orbit of the Moon) or in relative values $\delta T_B/T_B \leq 2/3 (M_m/M_\oplus)(R_B/R_m)^3 \approx 4 \cdot 10^{-8}$ (lines 3). The contribution of the attraction of the Moon becomes decisive for lunar cable systems but this situation must be considered separately.

4. Attraction by the Sun. A similar estimation for the gradient of the gravitational field of the Sun yields $\delta T_B/T_B \leq 2/3 (M_\odot/M_\oplus)(R_B/R_\odot)^3 \approx 2 \cdot 10^{-8}$ (lines 4), where $M_\odot$ and $R_\odot$ denote the mass of the Sun and the distance to it, respectively.

5. Relativistic effects. In the relativistic approximation [62] the change of tension yields $\delta T_B/T_B \sim v^2/c^2 \approx 7 \cdot 10^{-10}$ (lines 5), where v is the velocity of the orbital motion and c is the velocity of light.

6. Attraction of the cable by the satellite. The force of the gravitational attraction of the carrying satellite of mass m_B and characteristic size $2r_B$ on the cable can be estimated as $F \approx G m_B \rho/r_B$. At typical values for the orbital plane $m_B \sim 100\,\text{t}$, $r_B \sim 10\,\text{m}$ [65] the relation $F/T_B \sim 10^{-9}$ is close to the level of relativistic perturbations. Therefore in Fig. 1.2a the appropriate line 6 is not separately drawn from line 5. We notice that in cable systems, deployed from Moon and Phobos, the satellite carrying the cable is a natural celestial body and its attraction is of first order.

7. Finite thickness of the cable. The force acting on a cross section of the cable of finite thickness d_t in a Newtonian field differs from force of influence on a material line by the value $\delta T_B/\delta T_B \approx 3(d_t/R_B)^2/16$ (line 7). For a cable of thickness $d_t = 5\,\text{mm}$ this relation yields 10^{-13}.

1.3.2 Bending and friction forces in the cable

The second group of perturbations corresponds to additional forces, which arise in an imperfect elastic flexible cable and also due to internal friction.

8. Bending stiffness of the cable. Let us consider the cable as a bar with bending stiffness E_{bend}, simply supported in point B and not subjected to the action of external forces. The eigenfrequencies of its bending oscillations will be equal to $\Omega_n = k_n^2 l^{-2}(E_{bend}/\rho)^{1/2}$, where k_n are roots of the equation $\tan k = \tanh k$: $k_1 = 0; k_2 = 3.93, \ldots; k_n \approx \pi(n - 3/4)$ [33]. The rotation of the cable as a rigid bar corresponds to a zero root. Its shape remains straight. Elastic bending forces do not arise. The elastic cable hanging down from the satellite has in the first mode of oscillation a straight shape. In the orbital plane the modal frequency is $\omega_1 = \sqrt{3}\omega$, where ω is angular velocity of the corresponding orbital motion; the highest modal frequencies of the transverse oscillations are defined by the formula $\omega_n = \sqrt{1.5(n + 1)}\omega$ ($n = 2, 3, \ldots$) [60]. At same numbers n the oscillation modes of the bar and the flexible cable have the same number of nodes n, counted together with the point B, and admit direct comparison.

Starting from the general definition of eigenfrequencies and eigenmodes of oscillations of mechanical systems [33], the resulting forces at an identical bending of the cable-bar and the flexible cable relate as squares of the frequencies Ω_n and ω_n. As in the scale of perturbations everything is compared to the maximum tension T_B in the cable, the maximum level of the bending forces should be defined as $F \sim T_B\Omega_n^2\omega_n^{-2}$ or $F/T_B \sim 2/3\pi^4 n^{-1}(n - 3/4)^4(n + 1)^{-1}E_{bend}\rho^{-1}\omega^{-2}l^{-4}$. With this formula the comparison should begin with $n = 2$, as the shape corresponding to $n = 1$ is straight and $F = 0$, $\Omega_1 = 0$. Bending stiffness of the cable plaited from n circular fibres each of diameter d_1 at comparatively small longitudinal loadings, which is available in the considered case, can be estimated as total bending stiffness of all fibres $E_{bend} \approx E_M n\pi d_1^4/64$ [103], and then the relation $E_{bend}/\rho \approx E_M d_1^2/16\rho_M$ holds.

For a thickness of fibres $D_1 = 50\mu m$ the relation F/T_B for the mode $n = 2$ is $\sim 5 \cdot 10^{-13}$ for steel (full line 8) and $\sim 2 \cdot 10^{-12}$ for kevlar (dotted–dashed line 8). With the increase of the number of modes n the relation F/T_B grows $\sim n^2$ and for $n = 10$ yields $\sim 10^{-10}$ and $4 \cdot 10^{-10}$ for steel and kevlar, respectively (line 8'). Usually the highest modes of oscillations $n \gg 1$ have no importance for practical computation of the motion of TSS.

9. Residual deformations in the cable. If the cable was stored on a drum, a long time after exit in free space it will try to take the form of a circular arch of some radius $R \geq R_d$ (R_d denotes radius of the drum). The gravitational gradient straightens the cable and keeps it in the vertical state. Thus because of residual deformations in the cable the bending moment $M_{bend} = E_{bend}/R$ is created, which corresponds to the transversal force $F = E_{bend}(Rl)^{-1}$ acting at the end (E_{bend} is bend stiffness of a cable). For the ratio of force to tension

the following estimation

$$F/T_B \leq 2/3 E_b \rho^{-1} \omega^{-2} l^{-3} R_d^{-1} \sim E_M d_1^2 \rho_M^{-1} \omega^{-2} l^{-3} R_d^{-1}/24, \qquad (1.2)$$

holds, where d_1 is the diameter of the fibres in the cable. For the radius $R_d = 10\,cm$ of the drum and the diameter of fibres $d_1 = 50\,\mu m$ this ratio yields $3 \cdot 10^{-9}$ for a steel cable and $8 \cdot 10^{-9}$ for a cable of kevlar (full and dotted–dashed lines 9). The inclination of these lines is slightly changed due to the necessary increase of the dimensions of the drum for an increase of the diameter of the cable.

10. Internal friction in the cable. For an oscillating cable its fibres rub against each other which results in loss of energy and damping of eigenoscillations. The dissipation of energy takes place in the material of the fibres. Starting from general theory of oscillations, the relation of the forces of internal friction to restoring elastic forces may be estimated as δ/π (δ denotes logarithmic decrement of attenuation). The relation $\delta/\pi = \eta$ is called loss coefficient [64]. In comparison with the maximum tension the force of friction does not surpass this limit: $F/T_B \leq \eta$. For the data [63, 116, 131] for longitudinal oscillations $\eta \leq 0.1$ (line 10). The basic contribution to the losses comes from the friction between fibres of the cable. But it essentially depends on the tension in the cable, which defines the amount of contact between the fibres. Increasing the tension the coupling of the fibres increases and the loss coefficient in the cable decreases coming closer to the loss coefficient in a separately taken fibre [49]. For the data [97] the loss coefficient in various materials is $\eta \sim 10^{-3}$ (line 10′).

The energy of transverse oscillations of a cable connected to a satellite is dissipated mainly because of the interaction with longitudinal oscillations. The additional transversal force F_{fr} caused by internal friction due to bending estimated to the elastic transversal force is $F_{fr} \sim \eta F_{bend}$, which already has been considered and has been shown to be extremely small.

*) Reserve of strength. For comparison in Fig. 1.2b the fracture tension T_* is depicted. It exceeds the equilibrium tension T_B for a steel cable 330 times (dashed line) and for a cable made of kevlar 2600 times (dotted–dashed line).

1.3.3   Electromagnetic forces.

These arise as result of the interaction of the cable with the geomagnetic field and ionospheric plasma.

11. Electrostatic charge of a dielectric cable. At a height of $400\,km$ positively charged ions basically are represented by ions of oxygen O^+ with a concentration $n_i \sim 10^5\,cm^{-3}$ [48, 81]. Their thermal velocity $v_i \approx 1\,km\,s^{-1}$, corresponding to a temperature $\Theta_i \approx 1500\,K$, is considerably smaller than the velocity of the orbital motion $v_0 \approx 8\,km\,s^{-1}$ whereas the thermal velocity of the electrons $v_e \approx 200\,km\,s^{-1}$, which corresponds to the same temperature $\Theta_e = \Theta_i$, strongly exceeds the orbital velocity v_0. In this case the flow of electrons running against a noncharged cable exceeds the flow of ions by a

factor $10^{1.5}$, and the cable is charged negatively resulting in such a potential, for which the electronic and ionic flows are counterbalanced. The potential of charging the side of the cable, which is directed into the orbital motion, may be estimated with the formula $\varphi \approx -k\Theta_e e^{-1} \ln(v_e/v_0)$, where k is Boltzmann's constant, e is charge of an electron [9, 10], which is equal to -0.4 V. The potential on the back side of the cable should be, by approximate estimations, one order larger, because the impacts of ions in the "shadow" for a quickly moving cable occurs significantly less often than the direct bombardment of the forward moving surface. Similar phenomena are observed on dielectric surfaces of ordinary satellites [9].

The electrical field of a charged cable is shielded in plasma at the expense of decreasing the concentration of electrons near the cable. Scale of the area of shielding is defined by the Debye radius $D = \sqrt{\varepsilon_0 k \Theta_e n_e^{-1} e^{-2}}$ (n_e is the concentration of electrons). For a characteristic height of 400 km the temperature $\Theta_e \approx 1500$K and the concentration $n_e \approx 10^5$ cm^{-3} [48, 81] $D \approx 9$ mm. In the considered case the simple estimations on the basis [10, 61] give the size of area of shielding $D_{ec} \sim 6D$ and the general charge of the cable $q \sim 2\pi\varphi\varepsilon_0 l / \ln(D_{ec}/d_T)$, where $\varphi \sim -1$ V is the average over the surface of the cable of the value of the potential of charging; ε_0 is the dielectric permeability in vacuum. In the equatorial area at a height of 400 km an induction of the geomagnetic field $B \approx 2.6 \cdot 10^{-5}$ Tl, and the Lorentz force acting on a charged cable is equal $F = qv_0 B \approx 6 \cdot 10^{-8}$ N. This estimation is almost independent of the diameter of the cable (dotted–dashed line 11).

12. Resistance of the plasma to the motion of a dielectric cable. The force of resistance of the plasma to the motion of a charged cable is defined by the flow of ions, colliding with the cable and the deviated electrical field of the cable inside the area of shielding: $F \sim m_i n_i v_0^2 D_{ec} l$ (m_i and n_i are mass and concentration of ions), in this case $F \sim 2 \cdot 10^{-4}$ N (dotted–dashed line 12).

13. Induced charge in a conducting cable. The electromagnetic influence on a conducting cable, in general, differs from the influence on a dielectric cable. The superfluous electrons, accumulated on the conducting cable, under the action of the Lorentz force $F = ev_0 B \cos i_0$ move in the direction of one end of the cable (i_0 is the inclination of the orbit to the magnetic equator). If the motion occurs from West to East and the cable is deployed downwards to the Earth the concentration of electrons will grow towards the bottom end of the cable. The accumulation of electrons will be stopped when the electrical field created by them in the cable will reach the intensity $E_0 = Bv_0 \cos i_0$, at which Lorentz's force is counterbalanced by the electrical force Ee. On slightly inclined orbits $E_0 \approx 0.2$ V m^{-1} the potential of the bottom end of a conducting cable of length $l = 20$ km reaches in comparison to a dielectric cable the large value $\varphi_A \approx -4000$ V! Accordingly, the diameter of the area of shielding D_{ec} sharply grows. Simple estimations give in average along the cable the value $D_{ec} \sim 1.5$ m and the total charge of the cable $q \sim \pi\varphi_A\varepsilon_0 l / \ln(D_{ec}/d_T) \approx -3 \cdot 10^{-4}$ K. Lorentz's force acting on the cable will be $F = qv_0 B \sim 6 \cdot 10^{-5}$ N.

This estimation depends also weakly on the diameter of the cable (full line 13).

14. Resistance of plasma against the motion of a conducting cable. The formula given above remains valid also for this case, only the average effective diameter D_{ec} becomes greater because of the high potential of charging φ. This results in an estimate of $F \sim 5 \cdot 10^{-3}$ N (line 14).

15. Electrical current in the cable. The part of positively charged ions, getting inside the area of shielding, hits the metallic surface of the cable and pulls electrons out of it. The intensity of this flow increases together with the negative charge at the bottom end of the cable. The electrons, which get lost as the result of such impacts are filled at the expense of impacts at the weakly charged top end of the cable. This results in an electrical current in the cable. For the total value of the current it is possible to give an estimation on the basis [61] $I_B \approx d_T l e n_i \ln(D_{ec}/d_T) \sqrt{|\varphi_A| e/m_i}$, where n_i is the undisturbed concentration of ions in the plasma. The total force, acting on a cable with a current in a magnetic field, is estimated, as $F \approx 0.6 I_B B l$; the coefficient 0.6 arises because of the change (variation) of the current along the cable (charge flows down along the whole surface of the cable). By these estimations in a metallic cable with a diameter $d_T = 3$ mm a current $I_B \sim 1$ A will arise that will cause Ampere's force $F \sim 0.3$ N. Increasing the diameter of the cable will also increase this value (dashed line 15). The obtained force is relatively large and in this situation a detailed electrodynamical investigation will be necessary. However, one hardly will be able to deploy a bare metallic cable at those heights, where the concentration of electrons is large enough and the down flow of charge in the plasma is very intensive. For the practical use of the effect of the induced potential the conducting cable needs to be isolated and equipped with contact devices to the plasma on its ends, as it is supposed in the electromagnetical TSS.

1.3.4   Aerodynamic drag, solar radiation and impacts of micrometeorites

16. Aerodynamic resistance. For a density of air $\rho_a = 2.5 \cdot 10^{-12}$ kg m^{-3}, which is characteristic for a height of 400 km [107], the cable with a diameter $d_T = 1$ mm and a length of $l = 20$ km is exposed to the resistance $F \approx \rho_a v_0^2 d_T l \approx 3.2 \cdot 10^{-3}$ N (v_0 is the orbital velocity). The aerodynamic force changes proportionally to the diameter of the cable (line 16) and increases quickly if the height of flight decreases. For a tethered atmospheric probe the influence of aerodynamics becomes decisive.

17. Light pressure. The force of light pressure depends on the angle of the cable to the solar rays and on the reflecting ability of the cable. In any case, for a cable with circular cross-section it does not surpass $F \leq 4/3 p_s d_T l$ (line 17), where $p_s \approx 4.5 \cdot 10^{-6}$ N m^{-2} is the light pressure on an orbit around the Earth [46]. For a cable of diameter 1 mm, $F \leq 1.2 \cdot 10^{-4}$ N.

18. Heating of the cable by solar radiation. A much more essential influence

of solar radiation results in heating of the cable. From the data [84] follows that the extrema of thermal-mechanical loadings occur at those instances when the cable enters and leaves the shadow of the Earth, that is, when the periods of heating or cooling of the cable are reversed. They last only a short time ($\sim 1\,min$). In a steel cable of a diameter 0.9 mm the maximum force is $F_{max} \approx 0.04 T_B$, and in cable of same mass made of Kevlar of a diameter 2 mm the maximum force appears one order less: $F_{max} \approx 0.002 T_B$. The average thermal-mechanical loadings are defined by the coefficient of linear expansion of the cable α_T and by the difference of temperatures of the cable $\Delta\Theta$ on its sunny and shadowy parts of its orbit: $F/T_B \sim \alpha_T \Delta\Theta$. Results of [84] are given for a steel cable $\alpha_T = 2 \cdot 10^{-5}\,\mathrm{K}^{-1}$, $\Delta\Theta \approx 150\,\mathrm{K}$, $F/T_B \sim 3 \cdot 10^{-3}$ (full line 18), for a cable from kevlar $\alpha_T = -2.5 \cdot 10^{-6}\mathrm{K}^{-1}$, $\Delta\Theta \approx 100\,\mathrm{K}$, $F/T_B \sim 3 \cdot 10^{-4}$ (dotted–dashed line 18). The characteristic peak loadings are shown by lines 18′. It is necessary to note that such loadings arise only for rigidly attaching the cable to the basic satellite. The level of loadings sharply falls for the installation of a longitudinal damper at the point of attachment, which is necessary in many cases.

19. Impacts of micrometeorites. From the available scientific data on the distribution of the masses of particles of meteoric flow [48, 75] it follows that the basic pulse results from meteoric particles of the size of $30 - 200\,\mu m$. For a cable with a diameter of 2 mm and a length of 20 km about once a day a particle of size $50 - 100\,\mu m$ carrying on the average impulse $p \sim 8 \cdot 10^{-5}\,\mathrm{kg\,m\,s}^{-1}$ will strike. The transfer of this impulse to the cable results in its small oscillations, and the dynamic effect of this perturbation is responsible for estimating the maximum returning force $F \sim p\omega \approx 10^{-7}\,\mathrm{N}$. The thickness of the cable has no influence on this estimation (line 19). With a change of thickness only the average time between the impacts of particles of a given size changes. For a thin cable $d_T \sim 0.1\,\mathrm{mm}$ the impact of such a particle will be both its first and final, because it will break. Therefore the line 19 is not finished on the left border.

1.4 Methods of mathematical modelling

The triple, physical model, mathematical model and methods of its analysis form the basis of an "experiment on paper" (analytical analysis or calculation by hand) or a "computer experiment." The mathematical model occupies a central place in this triple. It integrates physical models of processes with mathematical methods of research and analysis and hence is decisive in all processes of theoretical research.

The realisation of "experiment on paper" has common laws with the realisation of full-sized experiments. For example, the success of theoretical research depends also in many respects on the clearness of the formulated problems and the concreteness of questions, the completeness of the scope of a problem. As a consequence, the complexity of the mathematical model frequently contradicts the depth of the analysis of specific questions. At the same time other environments of activity allow quickly and with minimum expenses to obtain the response to an inquiry. Fast changeability and easy repeatability of mathematical experiments make this method of mathematical modelling a highly effective tool of research. Sufficiently complete models of the dynamics of TSS created up to the present time and packages of their program realisation such as GTOSS, KKYHOOK, MODEL.3, and NEW1B allow one to judge the serviceability of these projects of TSS in the beginning of their development with high reliability [34]. Computational models, which are used for the verification of the serviceability of TSS and the exact prediction of its motion, take into account the highest possible properties of the real TSS.

The mathematical modelling also provides a unique opportunity for the analysis and development of ideas about specific laws and peculiarities of real processes in their causal-consequential interrelation on the basis of "thought-experiment," when only essential elements for the analysis of the investigated phenomenon are allocated and are kept in the mathematical model. Mathematical models used for such researches displaying only separate peculiarities of the process can be rather far away from the adequate description of real systems (for example Keplerian motion and real motion of a spacecraft on orbit around the Earth). But just these problems allow representation of laws of real processes to develop. Such problems have obtained the name of "model problems."

Problems of the mathematical description and analysis of models of TSS dynamics are essentially new in comparison with earlier investigated problems of Space flight dynamics. Therefore, researches of TSS dynamics pose problems both for the development of methods of modelling and of the mathematical analysis of the models. So-called problem models directed first of all to the development of methods of modelling are on the border between working computational models and questions: how to describe? and how to calculate? It is natural that this side constantly moves from simple to complex. For example, taking into account the motion of end bodies of TSS results in equations of dynamics where the equations of dynamics of a flexible string have boundary conditions depending on higher derivatives. As it is known, developed algorithms applying finite-differences methods for the numerical solution of the partial differential equations assume that the boundary conditions do not depend on higher derivatives. Therefore the solution of this problem is possible either on the basis of development of appropriate computational methods or by change of ways of construction of the model [108].

The review of works on the dynamics of TSS shows that all three types of models — computational models, model tasks and problem models have wide applications in the research of dynamics of TSS.

1.4.1 Basic model: Point masses connected by a massless string

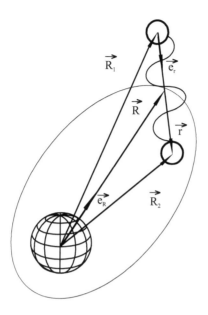

FIGURE 1.3
System of two tethered bodies.

The model of two point masses connected by a massless string (Fig. 1.3) is the basic problem model of TSS dynamics allowing to investigate laws of its motion. This model was the first model of research of TSS dynamics [86]. Even today it is widely used in both theoretical [50] and practical researches, for example, in investigations of processes of deployment (experiments SEDS-1,2 [39]) and retrieval of TSS. The equations of motion of a system of two point masses connected by a weightless string in a Newtonian field of forces are of the following form

$$m_1 \ddot{\vec{R}}_1 = -\frac{\mu m_1 \vec{R}_1}{R_1^3} + T_1 \vec{e_r} + \vec{F}_1,$$

$$m_2 \ddot{\vec{R}}_2 = -\frac{\mu m_2 \vec{R}_2}{R_2^3} - T_1 \vec{e_r} + \vec{F}_2, \tag{1.3}$$

where m_i are the masses of material points, $\vec{R}_i$ are their position-vectors with respect to the Newtonian attraction centre, $T_1 \vec{e_r}$ is the force acting along the connecting line (elastic force of the string), $\vec{e_r}$ is the unit vector directed along the connecting line, $\vec{F}_i$ is the total vector of other forces acting on the ith body ($i = 1, 2$), μ is the gravitational constant.

From (1.3) we obtain the equations of relative motion and the equations of motion of the mass centre of the tether

$$\ddot{\vec{r}} = \ddot{\vec{R}}_2 - \ddot{\vec{R}}_1 = -T\vec{e}_r + \vec{F}, \tag{1.4}$$

$$\ddot{\vec{R}} = -\frac{\mu\vec{R}}{R^3} + \vec{F}^*, \tag{1.5}$$

where $\vec{R} = \dfrac{\vec{R}_1 m_1 + \vec{R}_2 m_2}{M}$ is the position-vector of the mass centre of the tether about the attractive centre, $M = m_1 + m_2$,

$$T = T_1\frac{M}{m_1 m_2}, \quad \vec{F} = \frac{\vec{F}_2}{m_2} - \frac{\vec{F}_1}{m_2},$$

$$\vec{F}^* = (\vec{F}_1 + \vec{F}_2)/M + \vec{F}^*_{gr}, \tag{1.6}$$

$$\vec{F}_{gr} = \mu(\vec{R}_1/R_1^3 - \vec{R}_2/R_2^3),$$

$$\vec{F}^*_{gr} = \mu\vec{R}/R^3 - \frac{1}{m}\sum_{i=1}^{2}\mu m_i\vec{R}_i/R_i^3. \tag{1.7}$$

1.4.2 Model of TSS with massive string

1.4.2.1 Tether equations

More often the model of a flexible string is used as model of a cable. The position of a point s at time t is defined by the position-vector $R(s,t)$. The tension forces $T(s+ds,t)$ and $-T(s,t)$ act on an element of the string ds of density $\rho(S)$ from neighbouring elements. For the considered element Newton's equation of motion of its mass centre is

$$\rho(s)ds\frac{\partial^2\vec{R}}{\partial t^2} = \vec{T}(s+ds,t) - \vec{T}(s,t) - \frac{\mu\rho(s)ds\vec{R}}{R^3} + \vec{F}ds, \tag{1.8}$$

where $\vec{F}$ are external forces referred to the unit length of the string. From (1.8) we obtain

$$\rho\frac{\partial^2\vec{R}}{\partial t^2} = \frac{\partial\vec{T}}{\partial s} - \frac{\mu\rho\vec{R}}{R^3} + \vec{F}. \tag{1.9}$$

It is the usual form of the dynamical equations of a flexible string [76, 104]. By definition, the flexible string does not resist a bend and the force of its tension is always directed along a tangent to a line of a string

$$\vec{T} = T\vec{e}_s, \quad \vec{e}_s = \left(\frac{\partial\vec{R}}{\partial s}\right)\bigg/\left|\frac{\partial\vec{R}}{\partial s}\right|, \tag{1.10}$$

where $\vec{e}_s$ is unit vector of a tangent to a line of a string. The value of force of tension is defined by the law of an extensibility. Ordinarily the Hook's law of extensibility is used

$$T = E(\gamma - 1), \ \gamma = \left| \frac{\partial \vec{R}}{\partial s} \right|, \tag{1.11}$$

where E denotes module of elasticity of a string. At substitution (1.11), (1.10) in (1.9) vector equation in partial derivatives of a wave type turns out.

1.4.2.2 Satellite equations

The boundary conditions are defined by motion of end bodies. The equations of motion of the mass centres of end bodies look like (1.3)

$$m_1 \frac{d^2 \vec{R}_1}{dt^2} = -\frac{\mu m_1 \vec{R}_1}{R_1^3} + \vec{e}_{SA} T_A + \vec{F}_1, \tag{1.12}$$

$$m_2 \frac{d^2 \vec{R}_2}{dt^2} = -\frac{\mu m_2 \vec{R}_2}{R_2^3} + \vec{e}_{SB} T_B + \vec{F}_2,$$

where indices A and B designate the values of $\vec{e}_s$ and T in points of fastening of the cable accordingly to the first and to the second bodies. The locations of these points of fastening is defined by position-vectors $\vec{d}_A$ and $\vec{d}_B$ relative to the mass centres of the appropriate bodies. The dynamic equations of motion of the end bodies around their mass centres look like

$$\bar{\bar{\mathcal{J}}}_1 \dot{\vec{\omega}}_1 = -\vec{\omega}_1 \times \bar{\bar{\mathcal{J}}}_1 \vec{\omega}_1 + \vec{d}_A \times \vec{e}_{SA} T_A + \frac{3\mu}{R_1^3} \vec{e}_{R_1} \times \bar{\bar{\mathcal{J}}}_1 \vec{e}_{R_1} + \vec{M}_1,$$

$$\bar{\bar{\mathcal{J}}}_2 \dot{\vec{\omega}}_2 = -\vec{\omega}_2 \times \bar{\bar{\mathcal{J}}}_2 \vec{\omega}_2 + \vec{d}_B \times \vec{e}_{SB} T_B + \frac{3\mu}{R_2^3} \vec{e}_{R_2} \times \bar{\bar{\mathcal{J}}}_2 \vec{e}_{R_2} + \vec{M}_2, \tag{1.13}$$

where $\bar{\bar{\mathcal{J}}}_i$ denotes the tensor of inertia of an end body, $\vec{\omega}_i$ is the vector of its absolute angular velocity around the mass centre, $\vec{e}_{R_i} = \vec{R}_i / \left| \vec{R}_i \right|$, $\vec{M}$ is the moment of the other forces acting on i^{th} body. Then the boundary conditions of the motion of the string — the motion of points A and B — are defined by the following expressions

$$\ddot{\vec{R}}_A = \ddot{\vec{R}}_1 + \dot{\vec{\omega}}_1 \times \vec{d}_A + \vec{\omega}_1 \times (\vec{\omega}_1 \times \vec{d}_A),$$

$$\ddot{\vec{R}}_B = \ddot{\vec{R}}_2 + \dot{\vec{\omega}}_2 \times \vec{d}_B + \vec{\omega}_2 \times (\vec{\omega}_2 \times \vec{d}_B), \tag{1.14}$$

where $\vec{R}_A, \vec{R}_B$ are the position-vectors of points of fastening with respect to the Newtonian attractive centre. The creation of a stable finite difference algorithm for the numerical solution of the equations (1.9)–(1.14) represents a difficult problem. Therefore the transformation of the model to a variational problem and its solution by direct methods, for example, by Riesz's method, frequently is performed for simpler problems. If the approach described above for the derivation of the equations of motion can be related to a variational differential d'Alembert's principle, the integral principle of Hamilton–Ostrogradski

$$\delta \int_{t_1}^{t_2} (\delta A + \delta T) dt = 0$$

results in the variational statement of the problem. Here $\delta A = \delta V + \delta W$ denotes the virtual work of the potential δV and other δW forces, δT is the variation of the kinetic energy of the system. Use of this or other assumptions about the motion of the string and its lengthening, the completeness of the account of internal and external forces, result in the construction of various models of motion of TSS from linear up to essentially non-linear [13, 66, 79].

Wide application in practical calculations has also been obtained by the model of a chain of concentrated masses for a heavy cable [12, 42, 78]. The mechanical character of the discretisation of this model is obvious. As comparisons show, calculations with this model are well coordinated with calculations with the non-linear model with distributed parameters even in the case of a weakly tensed string [79].

Taking into account the bending and torsional stiffness of the cable, which could be important in the modelling of processes of deployment and retrieval of TSS, will influence the dynamics of the string and hence will create additional problems with ongoing discussions concerning the proper modelling.

1.5   Known results and some problems

As it is known, basic results on the dynamics of TSS as systems with distributed parameters are obtained for stationary or quasistatic modes of motion when the character of oscillations is linear. Sufficiently complete investigations of these modes of TSS motion are carried out in [25, 69].

The investigation of dynamics of systems with distributed parameters in a general case is possible only by methods of numerical integration and usually requires significant time expenses for realisation of calculations even on fast computers [102, 109, 110, 124]. We list some reasons why performing analysis of the dynamics of TSS as systems with distributed parameters is extremely complex and labourious:

- The boundary conditions of the continuous structure (the string) are defined by the dynamics of the end bodies and by the way of their attachment.

- The effect of external forces depends both on the motion of the cable and the bodies of the system.

- The dynamical characteristics of strings, especially the properties of extension and damping, are at present investigated only for linear modes of motion [11].

Generally speaking, the problem of construction of adequate mathematical models of TSS dynamics for essentially non-linear modes of motion, for example, for processes of retrieval and deployment of TSS, seems to be such a difficult problem that the creation of real TSS seems to be a problem of smaller complexity. In fact, taking into account the torsional stiffness of the cable and its elastic bending stiffness is essential for a cable stored in form of a band. For the analysis of the possibility of formation of loops, plastic deformations in the process of reeling and unreeling of the cable, the cable's heterogeneities, etc. form a problem of mathematical modelling which at present has not been clearly presented.

Thus, we can speak only about specification and improvement of models, about introduction of new characteristics, essential for the given mode of motion, about increase of degree of adequacy of models to real physical processes, about increase of accuracy and reliability of methods of calculations. Just in these directions investigations are conducted at the moment: [13, 31, 79, 108].

Existing models of TSS can be significantly improved both from an experimental and from a computational point of view. It seems to be very important to include the dynamics of the deployment system in the whole system dynamics during processes of deployment and retrieval. Until now the motion of the string from the point of its exit from the drum is described only in investigations of TSS. In more complex models, forces of resistance of the string at the point of its exit are given for deployment or retrieval. However, it would be necessary to simulate more complex effects, arising during unreeling of the string depending on the way it's arranged on the drum. In more sophisticated models of the system such peculiarities will play an important role and should be investigated.

Models of TSS with distributed parameters take into consideration the dynamics of the connecting string, the basic element of TSS. These models are much closer to the purpose of the adequate description of TSS dynamics than models of TSS as systems of connected rigid bodies. However, the basic method of analysis of non-linear interactions using these models is the method of numerical integration. Taking into account the complex models and complicated methods of their numerical integration it is difficultly to assume that the given approach will allow establishment of general laws of non-linear dynamics of TSS.

Problems of non-linear dynamics arise as basic problems in the investigation of the dynamics of space systems in different areas of their application. The tendency in the development of space systems, according to which, on the one hand, their overall size is considerably increased, the stiffness decreased and the inertial characteristics increased, and on the other hand, the accuracy requirements to fulfill the mission program become tougher, requires the solution of a series of fundamental problems of non-linear mechanics. Among such problems it is especially possible to select the problem of the influence of oscillations of masses of internal degrees of freedom on the dynamics of the system in a central field of forces and the problem of evolution of motion of extended systems with the allowance of coupling of relative and orbital motions. The indicated problems include problems of redistribution of energy in resonant modes, problems of stochastisation and synchronization of motions. The analysis of the present state of these problems and the history of development of Astronautics shows that in many cases it is difficult to foresee and predict effects of non-linear dynamics of systems beforehand. Therefore their investigation first of all requires the construction of a qualitative picture of non-linear dynamics, i.e., the development of ideas of basic regular properties and possible effects of dynamics of systems at non-linear interactions.

Let us consider, for example, questions, the solution of which requires creation of the TSS rotating on the orbit: How will the plane of TSS rotation change? How will its velocity of rotation around the mass centre change? Will resonant modes of motion be possible and what will be their effect? The solution of such questions is connected with the indicated problems of non-linear dynamics.

One basic peculiarity of TSS is the low stiffness of the connection of the bodies. By virtue of it, and also by virtue of the unidirectionality of the action of cables, modes of TSS motion with large frequencies of oscillation of internal degrees of freedom are possible, the character of which is essentially non-linear. In addition, these oscillations, despite of dissipation of their energy by internal friction, can permanently be excited in TSS motion at the expense, for example, of thermal shocks due to crossing of TSS the line of the terminator (borders where the TSS enters and leaves the Sun shadow of the Earth). Hence, investigation of the dynamics of rotational motion is connected to the solution of the problem of non-linear mechanics of the influence of essentially non-linear oscillations of bodies of internal degrees of freedom on the dynamics of systems in a central force field.

The review of publications on this problem shows that up to the present time there is fair amount of works on the dynamics of systems of connected bodies with oscillatory elements (see, for example, [2, 52, 68, 98, 111, 117]). However, the majority of investigations is carried out under the assumption of small amplitudes and quasistatic behaviour of oscillations of the system stipulated by finite stiffness of the connections. This assumption corresponds to linear oscillations of the system in the internal degrees of freedom, or in general, eliminates the effect of own elastic oscillations of the system from the consideration. For tasks in such a formulation there are well developed techniques of investigation at present available. Investigations of the dynamics of systems of the bodies for oscillations of internal degrees of freedom with large amplitude and the bodies losing connection, as well as techniques of investigation of dynamics of such systems, to the present time remain to be carried out.

The large extension of TSS stipulates an essential increase of forces and moments influential on its motion. The correctness of the consideration of TSS motions on the orbit of its mass centre within the framework of limited formulation of the task must each time be confirmed. At the same time, the research problem of long-term TSS rotation on the orbit requires the careful analysis of the influence both of environment, and properties of connection, as the long-term effect of even small environmental forces can result in significant deviations of the motion from the programmed one. Hence investigation of the dynamics of rotational TSS motion is connected to the solution of the problem of the evolution of the motion of extended systems on orbits around the Earth.

Just the cable system has served as the basis for the description of the gravity flyer idea [20]. However, problems of changes of aircraft attitude of extended systems had no direct practical importance earlier and many of its aspects remained uninvestigated until now. Hence, the analysis of the correlation of orbital and relative motions in the Newtonian field of forces is far from being completed. Further development (see [53, 74]) is needed for investigations of capabilities of mission control of systems of connected bodies in a Newtonian field of forces by the way of redistribution of the moment of momentum between orbital and relative motions by means of internal forces.

Let us consider the problem of possible stochastisation of motions. The area of investigation of chaotic motions of deterministic systems is one of the most recent areas of non-linear mechanics. Regular methods of analysis and more engineering type methods of investigation of applied problems are absent until now. It is natural in such a case to use the elementary model problem of TSS dynamics keeping in it the interaction between the internal degrees of freedom and the rotational motion of the system around its mass centre. We will consider the motion of a system of two material points in the plane of a circular orbit of the mass centre (Fig. 1.4) as such model with the assumption that the external force is caused only by the Newtonian field of forces. From (1.4) it is easy to obtain

$$\dot{L} = -\frac{3}{2}\frac{\mu}{R^3}r^2\sin 2\psi, \quad \dot{\psi} = \frac{L}{r^2} - \omega_0, \tag{1.15}$$

where L is the value of the specific moment of momentum of the relative motion of the tether, $L = |\vec{r} \times \dot{\vec{r}}|$, ψ is the angle between $\vec{r}$ and $\vec{R}$, $\omega_0 = \sqrt{\mu/R^3}$ denotes the constant angular velocity of the motion of the mass centre. In (1.15) the gravitational influences are taken into consideration assuming $r/R \ll 1$. We assume that the distance between the masses varies periodically, for example, $r = a + b\cos k(t - t_0)$, a, k and b are constants. In this case the model is very close to the model of two point masses connected by a linear spring. More detailed information about this will be given in Section 3.4. For $b/a \ll 1$ and $\dot{\psi}/\omega_0 \ll 1$ it is easy to see that according to the Kolmogorov-Arnold-Moser theory (see [15]) the trajectories of motion of the system are separated by invariant tori and will "forever" remain close to the undisturbed trajectories. However, at other ratios of parameters the conditions

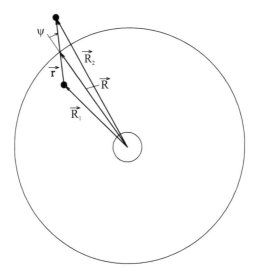

FIGURE 1.4
Two tethered bodies in orbit.

for stability of the undisturbed trajectories are violated and as a preliminary phase space analysis by Poincaré's point mapping method shows, the space gets a complex structure, where areas of resonant, conditionally periodic and chaotic trajectories occur.

Problems of TSS dynamics described in this section determine a circle of problems of the theoretical investigations considered in the book.

2

Equations of Motion of Space Tether Systems

2.1 Some remarks concerning the motion of TSS

There are typical situations in the dynamics of space systems when small
disturbances act on the system in addition to the main forces and moments
[121]. Such situations occur both for orbital motions and for many cases of
motion around the mass centre in the task of orientation and stabilisation of
the motion of space systems. In unperturbed motion due to the design of the
mechanical system the main regular properties of the motion are known. The
smallness of disturbances guarantees that on a small time interval the perturbed motion differs only little from the unperturbed one. Long-term effects
of perturbing forces can result in accumulation of disturbances in the motion
of a system and essentially change its characteristics in comparison with the
unperturbed motion. Therefore the main purpose of research of perturbed motions is the research of changes of characteristics of motion over a long time
interval and the determination of regular features of its evolution.

The disturbances are usually stipulated by the effect of the external force
fields on the system and depend on the motion of the system. Since the dynamics of space systems and the force fields are non-linear, the expressions of
the disturbances in generalized coordinates of motion of the system, in general,
are also non-linear and complicated, and the integration of these equations is
possible only in special cases. The research of dynamics of perturbed motion
by numerical methods, except for the well known shortages of these methods,
which are connected to the fact that only one particular trajectory is calculated, has one more characteristic property. In the presence of high-frequency
and low frequency oscillations in the system, the numerical research of the
perturbed motion of the system on a long-term time interval becomes prob-
lematic. High-frequency oscillations require a small integration step size that
causes an increase in the number of computations and appropriate errors. This
problem of the calculation in the dynamics of space systems has been known
for a long time. So, for example, it is known that in the tasks of calculation
of trajectories of motion of a satellite, in the investigation of the influence
of small disturbances the numerical integration in Cartesian coordinates is
useless.

The non-linear character of interaction is an essential aspect of the problem of investigation of perturbed motions of space systems. The solution of

the basic problem of investigation of perturbed motions of space systems is connected to the necessity to draw a qualitative picture of non-linear dynamics and to develop ideas about the main regular features and possible dynamic effects of systems for non-linear interactions. The problems of investigation of perturbed motion of space systems are tightly connected to the tasks of celestial mechanics. Just for the problem of perturbed Keplerian motion, the greatest scientists (mechanicians) created methods of research, which have constituted the basis of modern methods of non-linear mechanics. Here it is necessary to point out that the process of investigation of perturbed Keplerian motion consist of two steps: use of method of osculating elements (method of variation of parameters of Lagrange [46]) for the derivation of equations of perturbed motion and the use of one or another method of research of the obtained equations. The equations of perturbed Keplerian motion are characterized by the fact that the effects of disturbances are described by expressions containing small parameters and that the generalized coordinates of motion are divided into fast and slow variables. The different forms of the equations of perturbed Keplerian motion are the basis of research of orbital motions in celestial mechanics.

The classical formula of the method of osculating elements consists in the fixation of the forms of the first integrals and in the task of variation of the arbitrary constants of these integrals. Such an approach results in the derivation of equations of perturbed motion to bulky, frequently not acceptable schemes. Hence, schemes of derivation of equations of perturbed Keplerian motion, the first of which is based on evaluation of Lagrange's brackets for elliptical elements [112], and the second on geometrical constructions [46], turn out as excessively bulky and hamper the introduction of new variables, convenient for the particular task. The ideas of the method of osculating elements have found further development in the problems of the dynamics of space flight.

The Lurie scheme of derivation of perturbed Keplerian motion [45], based on the fixation of vectorial expressions of unperturbed motion, has considerably reduced the necessary quantity of transformations. The equations describing the perturbed motion of a rigid body about the mass centre (Euler-Puanso equations of perturbed motion [70]), derived on the basis of the extended method of osculating elements with the use of the theorem of change of moment of momentum, have allowed fundamental research of rotational motions of a satellite about its mass centre. Derivation of equations of perturbed motion of a free rigid body containing elastic and dissipative elements [36, 37], and the offered scheme of the method of averaging has allowed us to conduct research of a series of regular features of rigid body dynamics with mobile masses. In [71, 118, 119, 120] the dynamics of a viscous-elastic body in a Newtonian field of forces is considered. The equations of perturbed motion are derived and the scheme of the method of averaging is offered. On this basis the stationary motions (and their stability) are investigated. The list of publications, in which certain new results on regular features of non-linear dynamics are obtained, can be continued and essentially extended. Hence, it would be desirable to point out the original approach of research of regular properties of perturbed resonant motions [98]. The analysis of these works shows that the progress in knowledge of regular properties of non-linear dynamics of space systems first of all is connected to the derivation of new forms of equations of motion. The problem of derivation of equations of perturbed motion of a system including the selection of variables of motion, is in fact a task of mechanics. Mathematics supplies various methods of investigation of differential equations. Derivation of equations of motion for the task of dynamics is an informal process of coupling of a physical model of the investigated object with methods of research of the mathematical model. In all problems of dynamics, simplicity and clearness of the form of equations of motion is important and, consequently, to a considerable degree, the success of research is determined by proper selection of the variables describing the motion.

The proper choice of variables for the particular problem is in turn determined by the depth of understanding of regular properties and correlation of the considered motion. In the general statement of the problems of dynamics the successful choice of variables depends only on experience and intuition of the researcher. In problems of perturbed motion of systems the situation is different, since it is supposed that the perturbing effects are so small that the main regular properties of the unperturbed motion will not be broken, at least on a short time interval.

Therefore in problems of research of perturbed motion the choice of variables is realised on the basis of the analysis of regular properties of unperturbed motion. The general criteria for the choice of "evolutional variables" are stated in [18]. Generalizing them [121] it is possible to say that the extended method of osculating elements consists in the fixation of convenient, first of all mechanically clear, forms of unperturbed motion with their subsequent variation. Our knowledge of regular properties of the dynamics of a system is mapped into these forms. The sequence of such motion from the supposition of regular features of the dynamics to their fixation and refinement gives a powerful method of research of non-linear dynamics of systems in general.

2.2 Two point masses connected by a massless elastic string

The motion of a system of two point masses connected by an elastic massless flexible string in a central Newtonian force field (Fig. 1.3) is considered. The equations of motion of the considered system look like (1.3) or (1.4), (1.5).

We suppose from now on that the ratio of the length of the connection $r = |\vec{r}|$ to the distance from the centre of mass to the attracting centre R is small. Accurate in the order of $\dfrac{\mu}{R^2}\left(\dfrac{r}{R}\right)^2$ we obtain for orbital motion

$$\vec{F}^*_{gr} = \frac{\mu}{R^2}\left(\frac{r}{R}\right)^2 \frac{m_1 m_2}{M^2} \times$$

$$\left\{3\left(\vec{e}_r, \vec{e}_R\right)\vec{e}_r + \frac{3}{2}\left[1 - 5(\vec{e}_r, \vec{e}_R)^2\right]\vec{e}_R\right\}, \qquad (2.1)$$

and for attitude motion

$$\vec{F}_{gr} = \frac{\mu}{R^2}\frac{r}{R}\left\{-\vec{e}_r + 3\left(\vec{e}_r, \vec{e}_R\right)\vec{e}_R + 3\frac{m_1 m_2}{M}\frac{r}{R}\left[\left(\vec{e}_r, \vec{e}_R\right)\vec{e}_r + \right.\right.$$

$$\left.\left.\frac{1}{2}\left(1 - 5(\vec{e}_r, \vec{e}_R)^2\right)\vec{e}_R\right]\right\}, \qquad (2.2)$$

where $\vec{e}_R = \vec{R}/R$, $(\cdot, \cdot)$ designates the scalar product of vectors.

The expressions (2.1), and (2.2) are obtained by expansion of expressions (1.7) in series of r/R. Here, the following is taken into account: $\vec{R}_1 = \vec{R} - \vec{\rho}_1$, $\vec{R}_2 = \vec{R} + \vec{\rho}_2$, $\vec{r} = \vec{\rho}_2 - \vec{\rho}_1$, $m_2\vec{\rho}_2 + m_1\vec{\rho}_1 = 0$. $\vec{F}_{gr}$ is the gravitational force, which acts on the relative motion of TSS, $\vec{F}^*_{gr}$ is the force additional to the Newtonian gravitational force acting in the orbital motion of the TSS.

Hence, the value $\vec{F}^*_{gr}$ has the order of smallness $\left(\dfrac{r}{R}\right)^2$, and $\vec{F}_{gr} - \dfrac{r}{R}$.

Concerning the values of perturbation accelerations $\vec{F}$ and $\vec{F}^*$ we suppose their smallness in the sense that the kinetic energy of orbital motion of the tethered system and of attitude motions are essentially larger than the work of the appropriate perturbing forces on the considered time interval.

The relative motion of the tethered system if the perturbing accelerations are identically equal to zero ($\vec{F} \equiv 0$) we call unperturbed relative motion. Otherwise, if $\vec{F} \not\equiv 0$ we call the relative motion of the tethered system perturbed relative motion. Concepts of unperturbed and perturbed motion of the centre of mass are introduced similarly.

Equations (1.4) and (1.5) have the same form as the equation of motion of a material point in a central force field under the influence of perturbations. In the general case, such equations cannot be integrated analytically because general techniques for their investigation are not yet developed.

For the case of the Newtonian central force field (equation of motion of the mass centre of the tethered system (1.5)) the powerful method of osculating elements based on deriving the equations of perturbed Keplerian motion is developed in Celestial Mechanics.

However, the direct application of this method for the general case of a central force field is impossible. The reason is that the Keplerian motion belongs to a narrow class of periodic motions of a point in a central field of forces. Hence the explicit dependence between distance and angular variable is known and the derivation of the equations of perturbed Keplerian motion

[45] is based just on these properties. In the general case of a central field of forces the motion of a point is doubly periodical, and it is not clear at all [125] how to obtain the explicit dependence between distance and angular variable.

Thus, the problem of research of perturbed motion of the tethered system requires derivation of the equations of perturbed motion, i.e., development of a technique of representation of the equations of motion in a form suitable for research with the help of asymptotic methods of non-linear mechanics.

2.3 Unperturbed motion

The unperturbed motion of tethered systems is described by a pair of equations of motion of a point in the central force field following to (1.5), (1.6)

$$\ddot{\vec{r}} = -T\vec{e}_r, \quad \ddot{\vec{R}} = -\frac{\mu}{R^2}\vec{e}_R. \tag{2.3}$$

One can obtain equation (2.3) from equations (1.4), (1.5) if one ignores the disturbing accelerations.

The motion of the centre of mass is a Keplerian motion, the properties of which are known.

In the general case of a central force the trajectory of the unperturbed motion of points is a planar curve, the plane of which passes through the centre of force. The equations of motion in this plane in polar coordinates look like

$$\begin{aligned} \ddot{r} - r\dot{\phi}^2 &= -T, \\ r^2\dot{\phi} &= L, \end{aligned} \tag{2.4}$$

where L is the value of the constant moment of momentum referred to the mass of the point (the constant of areas).

If T is a function only of r, the energy integral

$$h = \frac{1}{2}\left(\dot{r}^2 + \frac{L^2}{r^2}\right) + \Pi(r), \tag{2.5}$$

exists, where h is the constant energy referred to the mass of the point, $\Pi(r) = \int T(r)dr$ and the solution of system (2.4) is reduced to quadratures

$$t - t_0 = \int \frac{dr}{\sqrt{f(r)}}, \quad \varphi = L\int \frac{dr}{r^2\sqrt{f(r)}} + \beta, \tag{2.6}$$

where

$$f(r) = 2h - 2\Pi - \frac{L^2}{r^2}, \tag{2.7}$$

t_0 and β are constants. However, it is possible to solve the quadratures, i.e., to present r and φ as functions of t or r as function of φ, only in some special cases.

Equation (2.5) describes the motion of the system with one degree of freedom. It is known [85, 35] that for systems with one degree of freedom in a force field two types of motions are possible: of libration (pendular) or limitation type (in the direction $\vec{r}$).

From now on we consider the motion of tethered systems as an elastic oscillator for which the change of variable r corresponds to librational motion, i.e., r in an unperturbed motion changes periodically between values r_1 and r_2, $r_1 < r_2$, which are the simple roots of the equation

$$f(r) = 0. \tag{2.8}$$

following to (2.7).

Then the change of r can be represented as

$$r = a - b\ \Phi(\omega(t)), \tag{2.9}$$

where $a = (r_1 + r_2)/2$ is the average value of the distance r, $b = (r_1 - r_2)/2$ is the amplitude of longitudinal oscillations, $\Phi(\omega)$ is a periodic function with respect to ω varied in the interval $[-1, 1]$, ω denotes phase of longitudinal oscillations, which is a function monotonically growing in time. The functions Φ and ω are connected by the equation

$$b^2 \left(\frac{d\Phi}{d\omega}\right)^2 \left(\frac{d\omega}{dt}\right)^2 = 2h - 2\Pi - \frac{L^2}{r^2}. \tag{2.10}$$

Equation (2.10) was obtained by substitution of (2.9) in (2.5).

Equation (2.10) allows if one of the functions Φ or ω is given to define the other. Usually [35, 70] two ways of definition of the functions Φ and ω are used. In the first way one defines $\Phi_1 = \cos\omega_1$, then

$$\dot{\omega}_1 = Q_1(r) = \left(\frac{f(r)}{(r - r_1)(r_2 - r)}\right)^{1/2} = \left(\frac{2h - 2\Pi - \dfrac{L^2}{r^2}}{b^2 \sin^2\omega_1}\right)^{1/2},$$

$$r = a - b\cos\omega_1. \tag{2.11}$$

As $r_1 \le r \le r_2$ and $f(r) = (r - r_1)(r_2 - r)g(r)$ the subradical expression is always larger than zero.

In the second way one determines $\omega_2 = \dfrac{2\pi}{\omega_{11}}(t - t_0)$, where ω_{11} is the period of longitudinal oscillations.

$$\omega_{11} = 2\int_{r_1}^{r_2} \frac{dr}{\sqrt{f(r)}} = 2\int_{0}^{\pi} \frac{d\omega_1}{Q_1(r)}. \tag{2.12}$$

Then

$$\frac{d\Phi_2}{d\omega_2} = \pm\frac{\omega_{11}}{b_2\pi}\sqrt{f(r)}, \quad r = a - b\Phi_2(\omega_2), \tag{2.13}$$

where the sign "+" corresponds to the decrease of r, and the sign "−" to the increase, and $\Phi_2(\omega_2)$ can be represented as Fourier series [85, 35]

$$\Phi_2(\omega_2) = B_0 + \sum_{n=1}^{\infty} B_n \cos n\omega_2, \tag{2.14}$$

where $\sum_{n=0}^{\infty} B_n = 1$.

It is possible to consider the angle φ as independent variable. Then the longitudinal oscillations are described by the equation of Binet [45]

$$L^2 v^2 \left(\frac{d^2 v}{d\varphi^2} + v^2\right) = T, \tag{2.15}$$

where $u = r^{-1}$, and the integral of energy (2.5) takes the form

$$h = \frac{L^2}{2}\left(\left(\frac{dv}{d\varphi}\right)^2 + v^2\right) + \Pi\left(\frac{1}{v}\right). \tag{2.16}$$

The change of u can be presented in the form, similar to (2.9). Thus the formulas determining the change of u in this representation are similar to (2.10)–(2.14). Taking into account that the amplitude of the longitudinal fluctuations b, average value of length a and the specific energy of the system h following to the low of preservation of energy are connected by the dependence

$$h = V(a + b) = V(a - b), \tag{2.17}$$

where $V = \Pi + L^2/(2r^2)$, it is possible to make the conclusion that the longitudinal oscillations in the general case are described by the formulas

$$\begin{cases} r & = r(h, L, \omega), \quad r(h, L, \omega + \pi_0) = r(h, L, \omega), \\ \dfrac{d\omega}{dt} & = Q(r, h, L) = \left(f(r)/\left(\dfrac{dr}{d\omega}\right)^2\right)^{1/2} \end{cases} \tag{2.18}$$

where π_0 is the period of oscillations of r with respect to the variable ω. (When $r = a + b$ or $r = a - b$ then $\dot{r} = 0$, and (2.17) follows from (2.5))

The change of the angle φ during one period of the longitudinal oscillations is equal

$$\omega_{12} = \Delta\varphi = 2L\int_{r_1}^{r_2} \frac{dr}{r^2\sqrt{f(r)}} = 2L\int_0^\pi \frac{d\omega_1}{r^2 Q_1(r)}. \tag{2.19}$$

The average frequencies of the longitudinal oscillations and rotations of a point around the attracting centre are equal, respectively

$$\mu_1 = \frac{1}{\omega_{11}}, \quad \mu_2 = \frac{\omega_{12}}{\omega_{11}2\pi}, \tag{2.20}$$

and, according to the theory of conditional-periodic motions [85, 35], the function $\sin \varphi$ can be expressed in the form of Fourier series as doubly-periodic function of variables

$$s_1 = \frac{2\pi}{\omega_{11}}(t - t_0),$$

$$s_2 = \beta + \frac{\omega_{12}}{\omega_{11}}(t - t_0). \qquad (2.21)$$

The motion becomes periodic in the case when μ_1/μ_2 is a rational value which occurs for

$$n_2\omega_{12} + 2n_1\pi = 0, \qquad (2.22)$$

i.e., ω_{12} and π are rationally commensurable. Then the period is equal $n_2\omega_{11}$.

Let us consider in detail the unperturbed motion (2.4) around the centre of mass of the tethered system of two material points with elastic massless connection, the elastic properties of which are described by Hook's law

$$T = \frac{C_m}{d}(r - d)\delta, \quad \delta = \begin{cases} 0, & r < d \\ 1, & r \geq d \end{cases}$$

$$C_m = C\frac{m_1 + m_2}{m_1 m_2}, \qquad (2.23)$$

where C is the coefficient of stiffness of the connection, d denotes the nominal length of the connection.

The function $f(r)$ looks like

$$f(r) = 2h - \delta\frac{C_m}{d}(r - d)^2 - \frac{L^2}{r^2}, \qquad (2.24)$$

and in this case the quadratures (2.6) are reduced to elliptic integrals. Their solution with respect to r obviously is not possible.

The area of possible motions of the tethered system is defined by the conditions $r \geq 0$, $f(r) \geq 0$. For $r \geq 0$ the function $f(r)$ has a single maximum in the point r_0:

$$\frac{1}{2}\frac{df}{dr} = -\frac{C_m}{d}(r_0 - d) + \frac{L^2}{r_0^3} = 0, \qquad (2.25)$$

where $r_0 (r_0 > d)$ corresponds to the equality of centrifugal and elastic forces, then for $2h = C_m(r_0 - d)^2/d = 2h_0$, $r_1 = r_2$ and the motion of the tethered system occurs on a circle of radius r_0. As $f(r) \to -\infty$ for $r \to 0$ and $r \to \infty$, the function $f(r)$ for $h > h_0$ has two simple real roots r_1 and r_2, $r_2 > r_0 > r_1$, which can be defined as roots of a polynomial of fourth degree $r^2 f(r)$. Consequently for $h > h_0$ the tethered system performs periodic longitudinal oscillations between the values r_1 and r_2. The phase portrait of the longitudinal oscillations for $C_m/d = 50\,\mathrm{s}^{-2}$ and $L/d^2 = 1\,\mathrm{s}^{-1}$ is presented in Fig. 2.1.

It is visible that the character of longitudinal oscillations of motion with

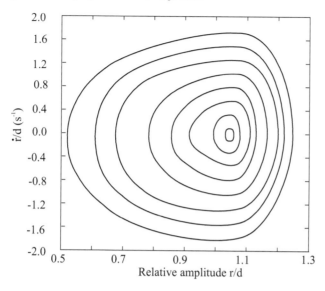

FIGURE 2.1
Phase portrait of longitudinal tether oscillations.

active (stretched tether) and non-active (tether is slack) connection can essentially differ.

First we consider the motion with active connection $r \geq d$. The condition of stretching the connection is equivalent to the inequality $f(d) \leq 0$, and it can be represented in form

$$\frac{L^2}{d^2} \geq 2h. \tag{2.26}$$

In the considered case the dependence between $r_1 = a - b$, $r_2 = a + b$ and h in accordance with (2.17) looks like

$$2h = \frac{C_m}{d}(r_2 - d)^2 + \frac{L^2}{r_2^2} = \frac{C_m}{d}(r_1 - d)^2 + \frac{L^2}{r_1^2}. \tag{2.27}$$

By virtue of this dependence

$$b^2 = a^2 - \sqrt{\frac{L^2 a}{\frac{C_m}{d}(a - d)}}, \tag{2.28}$$

and the expression for Q_1 (the formula (2.11) looks like

$$\dot{\omega}_1 = Q_1(r) = \left[\frac{C_m}{d} - \frac{L^2}{r_2^2 r_1^2 r^2}\left(2a(r + r_1) - r_1^2\right)\right]^{\frac{1}{2}},$$

$$r = a - b\cos\omega_1. \tag{2.29}$$

The character of the longitudinal oscillations reflects the most complete form of the function $\Phi_2(w_2)$ calculated through numerical integration of the formulas (2.12), (2.13) (Figs 2.2–2.4). Here we point out that the relation of the average frequencies μ_1 and μ_2, and also the function $\Phi_2(w_2)$ can be represented, depending only on two constant parameters of motion. In fact, as

$$f(r) = d^2 L^{*2} \left[\frac{2h}{d^2 L^{*2}} - \frac{C_m}{dL^{*2}}(z-1)^2\delta - \frac{1}{z^2} \right] = d^2 L^{*2} f^*(z),$$

where $z = \dfrac{r}{d}$, $L^* = \dfrac{L}{d^2}$,

$$\omega_{11} = \frac{2}{|L^*|} \int_{z_1}^{z_2} \frac{dz}{\sqrt{f^*(z)}} = \frac{\omega_{11}^*}{|L^*|}, \quad \omega_{12} = \frac{2L^*}{|L^*|} \int_{z_1}^{z_2} \frac{dz}{z^2 \sqrt{f^*(z)}} = \omega_{12}^*,$$

$$\mu_1 = \frac{|L^*|}{\omega_{11}^*}, \quad \mu_2 = \frac{\omega_{12}^* |L^*|}{\omega_{11}^* 2\pi}, \quad \frac{d\Phi_2}{dw_2} = \pm \frac{\omega_{11}^*}{b^* 2\pi} \sqrt{f^*(z)},$$

where $b^* = b/d$. Hence, the relation of the mean frequencies and the form of function $\Phi_2(w_2)$ depend only on two parameters

$$2h/(d^2 L^{*2}) = h^*, \quad C_m/(dL^{*2}) = c^*.$$

In Figs 2.2–2.4 the changes of the function $\Phi_2(w_2)$ within one period of oscillations depending on the value c^* are shown. This dependence is shown for longitudinal oscillations without losing connection with a small amplitude

$$z_1 = \frac{r_1}{d} = \frac{r_0}{d} - \frac{r_0 - d}{d}/100 = z_0 - (z_0 - 1)/100,$$

$$h^* \approx h_0^* = \frac{2h_0}{d^2 L^{*2}};$$

in Fig. 2.3 for longitudinal oscillations without losing connection with the almost greatest possible amplitude $z_1 = 1+(z_0-1)/1000$, $(h^* \approx 1)$; in Fig. 2.4 for longitudinal oscillations with losing connection $z_1 = 0.5$, $(h^* = 4)$. In Fig. 2.2 it is visible that for small amplitudes of oscillations $\Phi_2(w_2)$ practically coincides with $\cos w_2$, i.e., the "arithmetic mean" of longitudinal oscillations $(r_1 + r_2)/2$ practically coincides with the "integrated mean." Increasing the amplitude of oscillations (by decreasing the "stiffness" c^*, Fig. 2.3) the "integrated mean" $\Phi_2(w_2)$ deviates more and more from zero. For the motion with losing connection (Fig. 2.4) the deviations of the "integrated mean" $\Phi_2(w_2)$ from zero for the majority of values c^* grows, and with the increase of c^* the change of its sign takes place (Fig. 2.5).

In case of small amplitude of oscillations $b \ll a$ the longitudinal oscillations are close to harmonic ones. In fact, according to (2.28) and (2.29)

$$a = r_0 + \frac{\partial a}{\partial \varepsilon}\varepsilon + \ldots = r_0 + 2\frac{r_0 - d}{4r_0 - 3d}\left(\frac{b}{a}\right)^2 + \ldots, \tag{2.30}$$

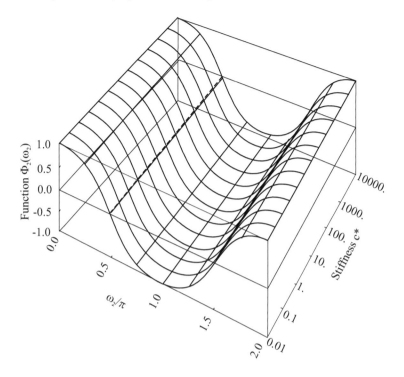

FIGURE 2.2
Function $\Phi_2(\omega_2)$ at longitudinal oscillations with small amplitude without loss of tension.

$$Q_1 = \sqrt{\frac{C_m}{d} + 3\frac{L^2}{r_0^4}} + \frac{1}{2}\frac{L^2}{r_0^4\sqrt{\frac{C_m}{d} + 3\frac{L^2}{r_0^4}}}\cos\omega_1\frac{b}{a} + \ldots .$$

Hence, in zero approximation with respect to the small parameter b/a longitudinal oscillations of the tether can be described by the formula

$$r = r_0 + b\cos k(t - t_0), \tag{2.31}$$

$$k = \sqrt{\frac{C_m}{d} + 3\frac{L^2}{r_0^4}}.$$

This representation of the variable r corresponds to the solution of the equations (2.8), $T = \dfrac{c_m}{d}(r - d)$, when the centrifugal accelerations L^2/r^3 are taken into account with the accuracy $\left(\dfrac{r - r_0}{r_0}\right)^2$. Accelerations which are not

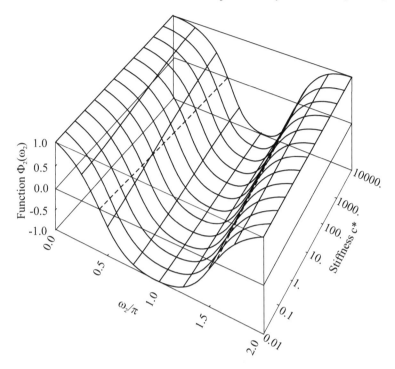

FIGURE 2.3
View of a function $\Phi_2(\omega_2)$ at longitudinal oscillations without losing connection with an almost maximal possible amplitude.

taken into account look like

$$F_{cm} = \frac{L^2}{r_0^3}\left[6\left(\frac{r-r_0}{r_0}\right)^2 - 10\left(\frac{r-r_0}{r_0}\right)^3 + \ldots\right] \tag{2.32}$$

The change of the angle φ in this case is equal

$$\varphi - \varphi_0 = L\int_{t_0}^{t}\frac{d\tau}{(r_0 - b\cos k(\tau - t_0))^2} =$$

$$= \frac{L}{k}\left[\frac{b\sin k(t-t_0)}{(r_0^2 - b^2)(r_0 - b\cos k(t-t_0))} + \right.$$

$$\left. +\frac{2r_0}{(r_0^2 - b^2)^{3/2}}\arctan\left\{\sqrt{\frac{r_0+b}{r_0-b}}\tan\frac{k(t-t_0)}{2}\right\}\right] \tag{2.33}$$

or if only members of first order of smallness relative to the small value b/a

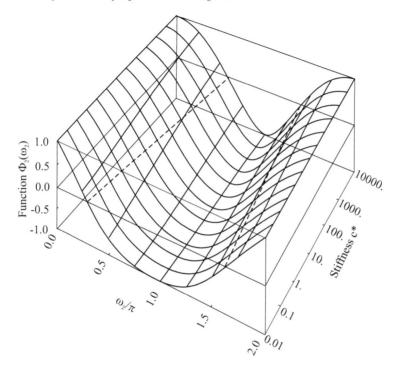

FIGURE 2.4
View of a function $\Phi_2(\omega_2)$ at longitudinal oscillations with loss of tension.

are kept in the integrand expression,

$$\varphi - \varphi_0 = \frac{L}{r_0^2}(t - t_0) - 2\frac{b}{r_0}\frac{L}{kr_0^2}\sin k(t - t_0) + \ldots \qquad (2.34)$$

At motion with loss of tension the dependence between $r_1 = a - b$, $r_2 = a + b$ and h, similar to (2.27), looks like

$$2h = \frac{C_m}{d}(r_2 - d)^2 + \frac{L^2}{r_2^2} = \frac{L^2}{r_1^2}. \qquad (2.35)$$

Hence, $r_1 = \sqrt{L^2/(2h)}$ and r_1 depend only on the value L and the velocity of activating connection as $2h = \dot{r}_n^2 + L^2/d^2$ where $\dot{r}_n$ is the rate of change of r at the moment of activating connection.

In free motion $r_1 \le r \le d$ (in motion with an unstrained cable $\delta = 0$) the quadratures (2.6) are integrated simply, and the change of r is described by the formulas

$$r = \sqrt{r_1^2 + 2h(t - \tau_0)^2},$$

$$\tau_0 = \sqrt{\frac{d^2 - r_1^2}{2h}} = \frac{d|\dot{r}_n|}{2h}, \qquad t \in [0, 2\tau_0], \qquad (2.36)$$

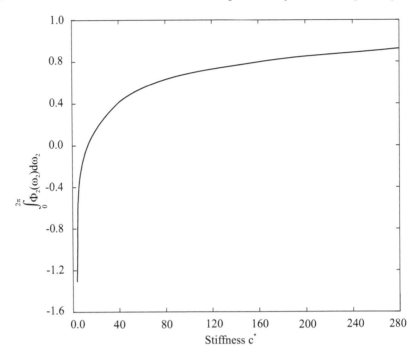

FIGURE 2.5
Change of $\int_0^{2\pi} \Phi_2(\omega_2)d\omega_2$ with increase of stiffness c^* for $z_1 = 0.5$.

where $2\tau_0$ is the time of the free motion of the tether within one period of longitudinal oscillations. The maximum value $2\tau_0$ is reached for $\dot{r}_n^2 = L^2/d^2$ and is equal to $d^2/|L|$, that is in inverse proportion to the angular velocity of rotation of the tether around the centre of mass at the moment of losing connection.

The change of the angle φ in free motion is described by formula

$$\varphi - \varphi_0 = \arctan\frac{2h}{L}(t - \tau_0) + \arctan\frac{2h}{L}\tau_0, \quad t \in [0, 2\tau_0]; \qquad (2.37)$$

and the differential of φ during free motion is equal to

$$\Delta\varphi = \varphi_k - \varphi_0 = 2\arctan\frac{2h}{L}\tau_0 = 2\arctan\frac{d|\dot{r}_n|}{L}. \qquad (2.38)$$

This increment grows with increase of the radial velocity of $|\dot{r}_n|$ and decreases with the increase of transversal velocity at the moment of losing connection. For $L \to 0$ and $|\dot{r}_n| \to \infty$ $\Delta\varphi$ tends to $\pm\pi$. From here it follows that the resonant ratio of low order 1:2 between the mean frequencies of longitudinal oscillations and rotation of the tether around the centre of mass, "the internal resonances" of the low order, can take place also at high stiffness of the connection.

Dependencies of the average frequencies of the longitudinal oscillations and the rotations of the tether on the value of the moment of momentum and energy of oscillations are represented in Fig. 2.6 and Fig. 2.7 for $C_m/d = 10\,\mathrm{s}^{-2}$.

The analysis of values of average frequencies shows that a resonant ratio μ_1 and μ_2 about 1:2, which is the resonance of "swing"-type, can take place only in cases of relatively low stiffness of the connection (parameter $c^* = c_m d^3/L^2$ is rather small or for a relatively large amplitude of longitudinal oscillations parameter h/h_0 is rather large).

Since the practical use of tethers supposes rather high stiffness of connection (parameter c^* is quite large) for reaching a resonant ratio of frequencies μ_1 and μ_2 the amplitude of oscillations, and hence the velocity for reaching connection should be so large that for the real system the connection simply would break. In Fig. 2.8 for $c^* = 100$ the change of the values μ_1, $\frac{1}{2}\mu_1$ and μ_2 depending on h/h_0 is shown. In Fig. 2.9 respective variations of the maximum and minimum values $r - (a + b)/d$, $(a - b)/d$, and the variation of a/d are shown.

Therefore, if the elastic properties of the string are described by Hooke's Law, the conclusion about the capability of the motion of such systems in resonance 1:2 represents most likely only theoretical interest, and investigation of this resonant mode has no practical importance.

In the mode of motion of small amplitude longitudinal oscillations $(b/a)^2 \ll 1$, with stretched connection the frequency of longitudinal oscillations by virtue of (2.25) is equal to

$$\frac{|L|}{r_0^2}\sqrt{\frac{r_0}{r_0 - d}} + 3.$$

Therefore, in this mode of motion the resonance 1:2 between frequencies generally can take place only in the limit for $r_0 \to \infty$.

The different character of the free motion of the system and the motion of the system with strained tether causes the impossibility of uniform description of motion. So, it obviously is not possible to represent the rate of change of the phase of oscillations in the description of r under the formula (2.11) by one function, which has no singular points. Taking into account (2.35) we obtain that for a motion with connection

$$\omega \in [\gamma, 2\pi - \gamma], \quad \gamma = arc\cos\frac{a - d}{b}, \tag{2.39}$$

$$Q_1 = \left\{\left[\frac{C_m}{d}(r_2 + r - 2d) - \frac{L^2}{r_2^2 r^2}(r_1 + r)\right]/(b - b\cos\omega)\right\}^{1/2}.$$

And in free motion $\omega \in [0, \gamma]$ and $\omega_2 = \dfrac{2\pi}{\omega_{11}}(t - t_0)$,

$$Q_1 = \left(\frac{L^2}{r_1^2 r^2}\frac{r_1 + r}{b(1 + \cos\omega)}\right)^{1/2}. \tag{2.40}$$

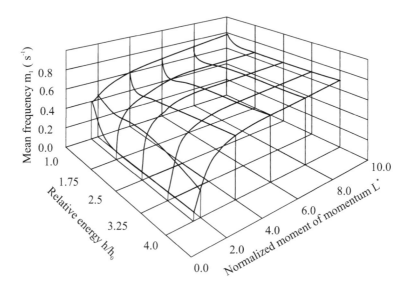

FIGURE 2.6
Relation of mean frequency of longitudinal oscillations on h and L^* at $c_m/d = 10\,\mathrm{s}^{-2}$.

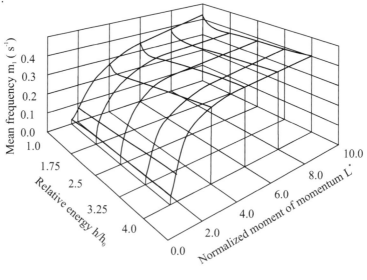

FIGURE 2.7
Relation of mean frequency of tether rotation on h and L^* at $c_m/d = 10\,\mathrm{s}^{-2}$.

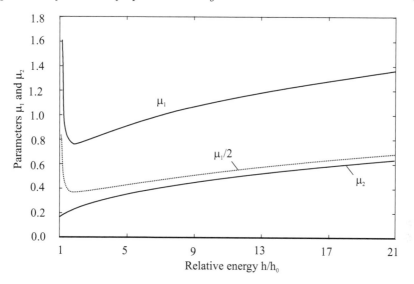

FIGURE 2.8
Dependence of $\mu_1, \mu_1/2, \mu_2$ on increasing energy of longitudinal oscillations, $c^* = 100$.

The change of the character of longitudinal oscillations is distinctly visible from the change of the function $\Phi(\omega_2)$, $\omega_2 = \dfrac{2\pi}{\omega_{11}}(t - t_0)$ (Figs 2.2–2.4). For a relatively stiff connection and certain parameters of motion if $(r_2 - d)/r_0 \ll 1$, it is possible to construct the approached solution for the motion with losing connection. Using the approximate solutions (2.31), (2.34) and solutions (2.36), (2.37) we obtain that the variation of r and of φ over one period of longitudinal oscillations is described by the formulas

$$r = r_0 - b\cos(kt + \gamma),$$

$$\varphi = \varphi_0 + \frac{L}{r_0^2}t + 2\frac{L}{r_0^3}\frac{b}{k}\left(\sin(kt + \gamma) - \sin\gamma\right)$$

and

$$r = \sqrt{r_1^2 + 2h(t - \tau_1)^2}, \qquad \tau_1 = \tau_0 + \frac{2\pi - 2\gamma}{k}, \qquad (2.41)$$

$$\varphi = \varphi_0 + \frac{L}{r_0^2}\frac{2\pi - 2\gamma}{k} - \frac{4L}{r_0^3}\frac{b}{k}\sin\gamma + \arctan\frac{2h}{L}(t - \tau_1) +$$

$$\arctan\frac{2h}{L}\tau_1, \quad t \in \left[\frac{2\pi - 2\gamma}{k}, 2\tau_0 + \frac{2\pi - 2\gamma}{k}\right],$$

where it is assumed that at the initial moment of time $t_0 = 0, r_n = d$, b and γ are defined from the equations

$$\begin{aligned} B\cos\gamma &= r_0 - d, \\ Bk\sin\gamma &= \dot{r}_n, \end{aligned} \qquad (2.42)$$

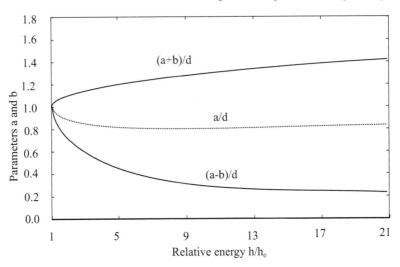

FIGURE 2.9
Appropriate variations of maximal, minimal and mean values of r.

where $\dot{r}_n$ is the velocity of coming on connection, and h and τ_0 are defined from formulas (2.35), (2.36).

Within the framework of the limited statement of the task, that is, the length of the tethered system is small compared to the radius of orbit, the regular properties of rotational TSS motion under the action of disturbances of different physical nature are studied. For different modes of motion the application of averaging is justified and the equations of first approximation are constructed. On the basis of these equations the analysis is conducted and the estimation of the evolution of the motion of a tethered system under the action of disturbances are constructed. The estimation of accuracy of the solutions of the equations of the first approximation are conducted by comparison of the obtained solutions with numerical solutions of the full equations.

2.4   Equations of perturbed motion

2.4.1   System with an elastically attached mass

Let us consider [121] the motion of a system, containing an elastically attached point mass (Fig. 2.10), whose motion is described by the equation

$$\ddot{\vec{r}} = -T(\bar{x}, \vec{r})\vec{e}_r + \varepsilon\vec{F}(\bar{x}, \bar{y}, \vec{r}, \dot{\vec{r}}), \qquad (2.43)$$

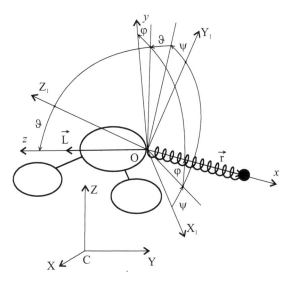

FIGURE 2.10
System, containing an elastically attached point mass.

where $\vec{r}$ is the position vector of the point mass about some pole O of the system, $\vec{e}_r = \vec{r}/r, r = |\vec{r}|$; $\bar{x}$, $\bar{y}$ are vectors of fast and slow variables, describing accordingly the motion of other parts of the system

$$\begin{aligned}
\dot{\bar{x}} &= \varepsilon \bar{X}(\bar{x}, \bar{y}, \vec{r}, \dot{\vec{r}}), \\
\dot{\bar{y}} &= \bar{y}(\bar{x}, \bar{y}, \vec{r}) + \varepsilon \bar{Y}(\bar{x}, \bar{y}, \vec{r}, \dot{\vec{r}}),
\end{aligned} \tag{2.44}$$

where ε is a small parameter.

Let us enter the right-handed coordinate systems: $CXYZ$ is an inertial system of coordinates, with respect to which the motion of the system is considered; $OX_1Y_1Z_1$ is the coordinate system connected to the motion of the system; $Oxyz$ is the coordinate system connected to the motion of the point mass, the axis Ox points in the direction of $\vec{r}$, the axis Oz has the direction of the vector of the moment of momentum of the point mass. Let the mutual orientation of systems $OXYZ$ and $Oxyz$ be determined by Euler angles ψ, θ and φ, which, respectively, represent precession, nutation and pure rotation angles.

Let us designate by $\vec{\omega}$ and $\vec{\omega}^*$, respectively, angular velocities $Oxyz$ relative to $OX_1Y_1Z_1$ and $OX_1Y_1Z_1$ relative to $CXYZ$. Then on the basis of the theorem of change of moment of momentum we obtain

$$\vec{L}' + (\vec{\omega} + \vec{\omega}^*) \times \vec{L} = \vec{r} \times \varepsilon \vec{F} = \vec{M}, \tag{2.45}$$

where $\vec{L} = \vec{r} \times \dot{\vec{r}} = L\vec{e}_3$, $\vec{e}_3$ is the unit vector of the axis Oz, the prime designates derivative of a vector with respect to time in the moving system of coordinates $Oxyz$.

The projections of equation (2.45) on the axes of the moving system of coordinates look like

$$\dot{L} = M_3, \quad (\omega_2 + \omega_2^*)L = M_1 = 0, \quad (\omega_1 + \omega_1^*)L = -M_2. \tag{2.46}$$

Here indexes 1, 2, 3 mean projections of the vector on the axes Ox, Oy, Oz respectively.

Using the known ratio between the components of angular velocity and the derivatives of Eulerian angles

$$
\begin{aligned}
\omega_1 &= \dot{\psi}\sin\theta\sin\varphi + \dot{\theta}\cos\varphi, \\
\omega_2 &= \dot{\psi}\sin\theta\cos\varphi - \dot{\theta}\sin\varphi, \\
\omega_3 &= \dot{\psi}\cos\theta + \dot{\varphi}
\end{aligned}
\tag{2.47}
$$

from the equations (2.46) we obtain [90]

$$
\begin{aligned}
\dot{\psi} &= \frac{\varepsilon r F_3 \sin\varphi}{L\sin\theta} + \cot\theta\,(\omega_{01}^*\sin\psi - \omega_{02}^*\cos\psi) - \omega_{03}^*, \\[2mm]
\dot{\theta} &= \frac{\varepsilon r F_3 \cos\varphi}{L} - \omega_{01}^*\cos\psi - \omega_{02}^*\sin\psi, \\[2mm]
\dot{L} &= \varepsilon r F_2,
\end{aligned}
\tag{2.48}
$$

where $\omega_{01}^*, \omega_{02}^*, \omega_{03}^*$ are projections of the vector ω^* on axes OX_1, OY_1, OZ_1 respectively.

The equations of change of the angle φ we find from equality

$$\vec{L} = L\vec{e}_3 = \vec{r}\times\dot{\vec{r}} = r^2(\omega_3 + \omega_3^*)\vec{e}_3. \tag{2.49}$$

Hence,

$$\dot{\varphi} = \frac{L}{r^2} - \dot{\psi}\cos\theta - \sin\theta\,(\omega_{01}^*\sin\psi - \omega_{02}^*\cos\psi) - \cos\theta\,\omega_{03}^*. \tag{2.50}$$

Taking into account that $\ddot{\vec{r}}\cdot\vec{e}_r = \ddot{r} - (\omega_3 + \omega_3^*)^2 r$, we obtain the equation of change of r:

$$\ddot{r} - \frac{L^2}{r^3} = -T(\bar{x}, \vec{r}) + \varepsilon F_1. \tag{2.51}$$

Thus, at the rather wide range of the angular velocity of motion of the system its equations can be considered as equations of motion of a system with an oscillatory element [123]:

$$\ddot{r} = -T_1(\bar{x}_1, r) + \varepsilon F_1(\bar{x}_1, \bar{y}_1, r, \dot{r}), \tag{2.52}$$

$$
\begin{aligned}
\dot{\bar{x}}_1 &= \varepsilon \bar{X}_1(\bar{x}_1, \bar{y}_1, r, \dot{r}), \\[2mm]
\dot{\bar{y}}_1 &= \bar{y}(\bar{x}_1, \bar{y}_1, r) + \varepsilon \bar{Y}_1(\bar{x}_1, \bar{y}_1, r, \dot{r}),
\end{aligned}
\tag{2.53}
$$

where $T_1 = T - L^2/r^3$, r is the generalised coordinate of the oscillatory element, $\bar{x}_1, \bar{y}_1$ are, respectively, vectors of fast and slow variables describing the motion of other parts of the system.

The motion of an oscillatory element is close to the motion of a conservative system with one degree of freedom. Therefore in unperturbed motion the solution of equations (2.52) is reduced to quadratures. In particular cases if these quadratures are solvable in explicit form the construction of the general solution of the equation of motion of an oscillatory element may be performed, which does not contain implicitly preset functions. The scheme of derivation of the equations of perturbed motion is rather simple [82]. Here we consider cases for which to construct the general solution of the equation of motion of an oscillatory element which does not contain implicitly specified functions, is not possible.

Introduction of "action-angle" variables [15] formally completely solves the problem of derivation of the equations of perturbed motion of systems, the equations motion of which are close to integrable Hamiltonian systems. However, the practical use of these variables has not obtained wide application, as it is connected to the solution of a number of rather complex problems, and the performance of generalised coordinates and velocities through "action-angle" variables is actually connected to the use of infinite Fourier series.

For systems, close to conservative systems with one degree of freedom, there is [90] a technique of construction of first and higher approximations, which do not require the solution of equations of unperturbed motion in explicit form. The absence in this technique of an explicit form of equations of perturbed motion results in a number of problems of excessively bulky transformations and evaluations, and also hampers the choice of the variables describing the motion convenient for the particular task.

The complexities connected to the integration of implicitly preset functions, in this technique, are overcome through averaging along the generating solution. However, in many cases this operation can not be carried out in analytical form and, therefore, it is required to be conducted in each step of a numerical integration. The technique offered below allows in many cases to simplify the process of derivation of the equations of first approximation for such systems, and essentially also to expand the class of systems investigated by the method of averaging.

In unperturbed motion $\varepsilon = 0$, $\bar{x}_1$ is constant, and the equation (2.52) has the first integral

$$h = \frac{1}{2}\dot{r}^2 + \Pi(\bar{x}_1, r), \quad \Pi = \int T_1(\bar{x}_1, r)dr, \tag{2.54}$$

where h is a constant, and the general solution is given by quadrature

$$t - t_0 = \int \frac{dr}{\sqrt{f(\bar{x}_1, r)}}, \quad f(\bar{x}_1, r) = 2h - 2\Pi(\bar{x}_1, r). \tag{2.55}$$

Similarly to the analysis given above of the unperturbed motion (2.5)–

(2.18) the variation of r can be presented in the way

$$r = a - b\,\Phi(w(t)), \quad a = (r_1 + r_2)/2, \quad b = (r_1 - r_2)/2,$$

where r_1, r_2 are simple solutions of equation $f(\bar{x}, r) = 0$, $r_1 < r_2$, $\Phi(w)$ is periodic in the function w which is varied on an interval $[-1, 1]$ and w is the phase of oscillations and a monotonically increasing function in time. The functions Φ and w are connected by the equation

$$b^2 (d\Phi/dw)^2 (dw/dt)^2 = 2h - 2\Pi. \tag{2.56}$$

If we accept $\Phi(\cdot) = \cos(\cdot)$ then

$$\dot{w} = Q_1 = \left(\frac{f(\bar{x}_1, r)}{(r - r_1)(r_2 - r)} \right)^{1/2} = \left(\frac{2h - 2\Pi}{b^2 \sin^2 w} \right)^{1/2}, \tag{2.57}$$

$$r = a - b\cos w_1.$$

If we accept that $w = (2\pi/\omega_{11})(t - t_0)$, where

$$\omega_{11} = 2 \int_{r_1}^{r_2} \frac{dr}{\sqrt{f(\bar{x}_1, r)}} = 2 \int_{o}^{\pi} \frac{dw}{Q_1(r)} \tag{2.58}$$

is the oscillation period, the vibration mode is described by the function $\Phi(\cdot)$:

$$\frac{d\Phi}{dw} = \pm \frac{\omega_{11}}{2\pi b} \sqrt{f(\bar{x}_1, r)}, \quad r = a - b\Phi(w), \tag{2.59}$$

where the sign "+" corresponds to the decrease of r and the "−" sign to its increase.

Since the oscillation amplitude is b, the mean value a and the energy constant h are connected by the relation

$$h = \Pi(\bar{x}_1, a + b) = \Pi(\bar{x}_1, a - b), \tag{2.60}$$

then in general the oscillation of the element can be described by the formula

$$\begin{aligned}
r &= r(\bar{x}_1, h, w), \quad r(\bar{x}_1, h, w + \pi_0) = r(\bar{x}_1, h, w), \\
\dot{r} &= (\partial r/\partial w)\dot{w}, \tag{2.61} \\
\dot{w} &= Q(\bar{x}_1, h, r) = \left[f(\bar{x}_1, r)/(\partial r/\partial w)^2 \right]^{1/2},
\end{aligned}$$

where π_0 is the oscillation period of r depending on w.

Hence, the oscillations of the element are characterized by the oscillation amplitude and the phase, and in general are described by the formulas (2.61). Let us consider h and w as new variables, and relations (2.61) as the formulas of replacement of variables. Differentiating (2.54) with respect to time by virtue of (2.52), we obtain

$$\dot{h} = Q(\partial r/\partial w)\varepsilon F_r + (\partial \Pi/\partial \bar{x}_1)\dot{\bar{x}}_1. \tag{2.62}$$

The second equation in (2.61) gives

$$\dot{w} = Q - \left(\dot{h}\partial r/\partial h + \dot{\bar{x}}_1 \partial r/\partial \bar{x}_1\right)(\partial r/\partial w)^{-1}. \tag{2.63}$$

Therefore, the general set of the equations of perturbed motion of the oscillatory element is determined by equations (2.62), (2.63).

From the derivation of the equations it is visible that any other constant of unperturbed motion describing the oscillation frequency of the element may be chosen instead of h. In particular, it can be some function of r_1, r_2. The connection between r_1, r_2, h and $\bar{x}$ is determined by relation (2.60). Therefore,

$$\dot{r}_i = \left[\varepsilon F_r Q \frac{\partial r}{\partial w} + \dot{\bar{x}}_1 \left(\frac{\partial \Pi(\bar{x}_1, r)}{\partial \bar{x}_1} - \frac{\partial \Pi(\bar{x}_1, r_i)}{\partial \bar{x}_1}\right)\right]\left(\frac{\partial \Pi(\bar{x}_1, r_i)}{\partial r_i}\right)^{-1}. \tag{2.64}$$

In the representation of r in the form $r = a - b\cos w$, where w in unperturbed motion is determined by equation (2.57), the equations of oscillation of the element have the form

$$\dot{b} = \left\{\varepsilon Q_1 b F_r \sin w_1 + \dot{\bar{x}}_1 \left[\frac{\partial \Pi(\dot{\bar{x}}_1, r)}{\partial \bar{x}_1} - \frac{\partial \Pi(\bar{x}_1, r_2)}{\partial \bar{x}_1} - \right.\right.$$

$$\left.\left.\frac{\partial a}{\partial \bar{x}_1}\frac{\partial \Pi(\bar{x}_1, r_2)}{\partial r_2}\right]\right\}\left\{\frac{\partial \Pi(\bar{x}_1, r_2)}{\partial r_2}\left(1 + \frac{\partial a}{\partial b}\right)\right\}^{-1} \tag{2.65}$$

$$\dot{w} = Q + \left(\dot{b}\cos w_1 - \frac{\partial a}{\partial \bar{x}_1}\dot{\bar{x}}_1 - \frac{\partial a}{\partial b}\dot{b}\right)/(b\sin w_1).$$

This form of the equations of the oscillating element is convenient because the oscillations of the element are described by trigonometrical functions. At the same time the dependence Q_1 on w in the general case limits the effective application of the operators of averaging actually to the case of one fast variable w in the motion of the system.

In the representation $r = a - b\Phi(w)$ where $w = (2\pi/\omega_{11})(t - t_0)$, ω_{11} is the oscillation period of the element in unperturbed motion, the equations of oscillations of the element take the form

$$\dot{b} = \left\{\varepsilon\frac{2\pi b F_r}{\omega_{11}}\frac{\partial \Phi}{\partial w} + \bar{x}_1\left[\frac{\partial \Pi(\bar{x}_1, r)}{\partial \bar{x}_1} - \frac{\partial \Pi(\bar{x}_1, r_2)}{\partial \bar{x}_1} - \right.\right.$$

$$\left.\left.\frac{\partial a}{\partial \bar{x}_1}\frac{\partial \Pi(\bar{x}_1, r_2)}{\partial r_2}\right]\right\}\left\{\frac{\partial \Pi(\bar{x}_1, r_2)}{\partial r_2}\left(1 + \frac{\partial a}{\partial b}\right)\right\}^{-1}, \tag{2.66}$$

$$\dot{w} = \frac{2\pi}{\omega_{11}}(t - t_0) - \left(\dot{b}\ \Phi(w) - \frac{\partial a}{\partial \bar{x}_1}\dot{\bar{x}}_1 - \frac{\partial a}{\partial b}\dot{b}\right)/\left(\frac{b\ \partial\Phi(w)}{\partial w}\right).$$

2.4.2 Motion of a mass point in the central force field

Let us consider [88] equations of motion of a mass point in the form (1.4), (1.5)

$$\ddot{\vec{r}} = -T\vec{e}_r + \vec{F}, \qquad (2.67)$$

where $\vec{r}$ is the position vector of the point about the centre of force, $\vec{e}_r = \vec{r}/r$, $r = |r|$, $-T\vec{e}_r$ is the acceleration of the point by the central force, $\vec{F}$ is the vector of perturbing accelerations.

Let us introduce the right coordinate systems (Fig.2.11): $DXYZ$ is the non-rotating coordinate system with origin in the centre of force D, $Dxyz$ is the moving coordinate system, the axis Dz points in the direction of the vector of moment of momentum, the axis Dx has the direction of the vector $\vec{r}$. The mutual orientation of the systems $DXYZ$ and $Dxyz$ is determined by Euler's angles ψ, θ and φ.

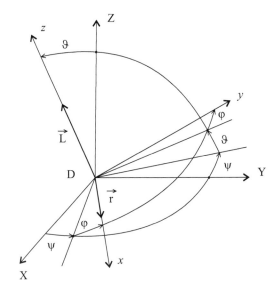

FIGURE 2.11
Euler's angles.

According to the above-stated technique (2.45)–(2.51) on the basis of the theorem of change of moment of momentum we obtain the first group of equations of perturbed motion

$$\dot{\psi} = \frac{rF_3 \sin \varphi}{L \sin \theta},$$

$$\dot{\theta} = \frac{rF_3 \cos \varphi}{L},$$

$$\dot{L} = rF_2,$$ \qquad (2.68)

$$\dot{\varphi} = \frac{L}{r^2} - \dot{\psi} \cos \theta$$

and the equation of radial oscillations

$$\ddot{r} - \frac{L^2}{r^3} = -T + F_1. \qquad (2.69)$$

The equations (2.68), (2.69) are valid for the arbitrary T. From now on we consider that the central force is such that the unperturbed motion of the point is described by formulas (2.4)–(2.18). Then according to the described technique the equations of perturbed radial oscillations in general take the form

$$\dot{h} = Q \frac{\partial r}{\partial \omega} F_1 + \frac{\dot{L}L}{r^2}, \quad \dot{\omega} = Q - \left(\dot{h} \frac{\partial r}{\partial h} + \dot{L} \frac{\partial r}{\partial L} \right) / \frac{\partial r}{\partial \omega},$$ \qquad (2.70)

$$r = r(h, L, \omega), \quad Q = Q(h, L, \omega).$$

The system of equations (2.68), (2.70) of perturbed motion in general case contains two fast variables ω and φ, corresponding to the phase of oscillations of the distance from the point up to the centre of the force and the angle of pure rotation.

As the connection between r_1, r_2, h and L is given by the formula (2.17), then

$$\dot{r}_i = \left[Q \frac{\partial r}{\partial \omega} F_1 + L\dot{L} \left(\frac{1}{r^2} - \frac{1}{r_i^2} \right) \right] / \frac{\partial V(r_i)}{\partial r_i}, \quad i = 1, 2, \qquad (2.71)$$

and in particular,

$$\dot{a} = Q \frac{\partial r}{\partial \omega} F_1 \frac{1}{2} \left(1 / \frac{\partial V(r_2)}{\partial r_2} + 1 / \frac{\partial V(r_1)}{\partial r_1} \right) + \frac{L\dot{L}}{2r^2} \left[(r_1^2 - r^2) / \right.$$

$$\left. \left(r_1^2 \frac{\partial V(r_1)}{\partial r_1} \right) + (r_2^2 - r^2) / \left(r_2^2 \frac{\partial V(r_2)}{\partial r_2} \right) \right], \qquad (2.72)$$

$$\dot{b} = Q \frac{\partial r}{\partial \omega} F_1 \frac{1}{2} \left(1 / \frac{\partial V(r_2)}{\partial r_2} - 1 / \frac{\partial V(r_1)}{\partial r_1} \right) + \frac{L\dot{L}}{2r^2} \left[(r_1^2 - r^2) / \right.$$

$$\left. \left(r_2^2 \frac{\partial V(r_2)}{\partial r_2} \right) - (r_1^2 - r^2) / \left(r_1^2 \frac{\partial V(r_1)}{\partial r_1} \right) \right]. \qquad (2.73)$$

For the introduction of r by $r = a - b \cos \omega_1$ where ω_1 in unperturbed motion is determined by the equation (2.11), taking b and ω_1 as new variables, we may write the equations describing radial oscillations by

$$\dot{b} = \left\{ Q_1 b F_1 \sin \omega_1 - \dot{L} \left[L \left(\frac{1}{r_2} - \frac{1}{r^2} \right) + \frac{\partial a}{\partial L} \frac{\partial V(r_2)}{\partial r_2} \right] \right\} \times$$

$$\left\{ \frac{\partial V(r_2)}{\partial r_2} \left(1 + \frac{\partial a}{\partial b} \right) \right\}^{-1}, \tag{2.74}$$

$$\dot{\omega}_1 = Q_1 + \left(\dot{b} \cos \omega_1 - \dot{b} \frac{\partial a}{\partial b} - \frac{\partial a}{\partial L} \dot{L} \right) / b \sin \omega_1.$$

For the representation $r = a - b \, \Phi_2(\omega_2)$, where $\omega_2 = \frac{2\pi}{\omega_{11}}(t - t_0)$, ω_{11} is the period of radial oscillations in unperturbed motion, the equations describing the radial oscillations look like

$$\dot{b} = \frac{-\dfrac{2\pi}{\omega_{11}} b F_1 \dfrac{d\Phi_2}{d\omega_2} - \dot{L} \left[L \left(\dfrac{1}{(a+b)^2} - \dfrac{1}{r^2} \right) + \dfrac{\partial a}{\partial L} \dfrac{\partial V(a+b)}{\partial a} \right]}{\dfrac{\partial V(a+b)}{\partial a} \left(1 + \dfrac{\partial a}{\partial b} \right)},$$

$$\dot{\omega}_2 = \frac{2\pi}{\omega_{11}} - \left(\dot{b}\Phi_2(\omega_2) - \dot{b}\frac{\partial a}{\partial b} - \frac{\partial a}{\partial L} \dot{L} \right) \Big/ \left(b\frac{d\Phi_2}{d\omega_2} \right), \tag{2.75}$$

where functions $\Phi(\cdot)$ and ω_{11} do not change in unperturbed motion, Φ_2, ω_{11} and $\dfrac{d\Phi_2}{dw_2}$ are defined from (2.12), (2.13) for initial values of the parameters of motion h, L.

For the description of radial oscillations of the point in unperturbed motion by Binet's equation (2.15) and by the integral of energy (2.16) the algorithm of deriving the equations of perturbed motion obviously does not change. Therefore considering the dimensionless value φ^0 (angle of pure rotation in unperturbed motion) as the independent variable and using the representation $\dfrac{1}{r} = v = a_v - b_v \Phi_3(w_3)$, where

$$w_3 = \frac{2\pi}{\omega_{12}}(\varphi^0 - \varphi_0^0),$$

ω_{12} denotes the period of longitudinal oscillations in new "time" φ^0 (2.19), and Φ_3 is defined similarly Φ_2 (2.13)

$$\frac{d\Phi_3}{dw_3} = \pm \frac{\omega_{12}}{b_v 2\pi} \sqrt{\frac{2h}{L^2} - v^2 - \frac{2}{L^2} \Pi \left(\frac{1}{v} \right)},$$

we obtain the equations of perturbed motion of the point in elastic connection in the following form

$$\frac{d\psi}{d\varphi^0} = \frac{F_3 \sin\varphi}{v^3 L^2 \sin\theta}, \quad \frac{d\theta}{d\varphi^0} = \frac{F_3 \cos\varphi}{v^3 L^2},$$

$$\frac{dL}{d\varphi^0} = \frac{F_2}{v^3 L}, \quad \frac{d\alpha}{d\varphi^0} = -\frac{d\psi}{d\varphi^0}\cos\theta, \tag{2.76}$$

$$\frac{db_v}{d\varphi^0} = \frac{\dfrac{1}{v^2}\dfrac{2\pi}{\omega_{12}} F_1 b_v \dfrac{d\Phi_3}{d\omega_3} - \dfrac{dL}{d\varphi^0}\left[L\left(v_2^2 - v^2\right) + \dfrac{\partial a_v}{\partial L}\dfrac{\partial V(v_2)}{\partial v_2}\right]}{\dfrac{\partial V(v_2)}{\partial v_2}\left(1 + \dfrac{\partial a_v}{\partial b_v}\right)},$$

$$\frac{d\omega_3}{d\varphi^0} = \frac{2\pi}{\omega_{12}} - \left[\frac{db_v}{d\varphi^0}\left(\Phi_3(\omega_3) - \frac{\partial a_v}{\partial b_v}\right) - \frac{\partial a_v}{\partial L}\frac{dL}{d\varphi^0}\right] \Big/ \left(b_v \frac{d\Phi_3}{d\omega_3}\right),$$

where $\varphi = \varphi^0 + \alpha$, and the dependence between the new independent variable φ^0 and time t is determined by the ratio

$$\frac{d\varphi^0}{dt} = Lv^2.$$

The equations (2.76) differ from earlier derived ones by the applicability to them of algorithms of averaging developed for the independent rotary systems [51].

Let the perturbing accelerations have a force function, i.e., such a perturbing function U exists that the projections of the perturbing acceleration on the axes of the non-rotating system of coordinates $DXYZ$ are defined by the formulas

$$F_x = \frac{\partial U}{\partial X}, \quad F_y = \frac{\partial U}{\partial Y}, \quad F_z = \frac{\partial U}{\partial Z}.$$

In this case the projections of the perturbing acceleration on the axes of the moving system of coordinates look like

$$F_1 = \frac{\partial U}{\partial r}, \quad F_2 = \frac{1}{r}\frac{\partial U}{\partial \varphi}, \tag{2.77}$$

$$F_3 = \frac{1}{r\sin\varphi}\frac{\partial U}{\partial \theta} = \frac{1}{r\cos\varphi\sin\theta}\left(\frac{\partial U}{\partial\varphi}\cos\theta - \frac{\partial U}{\partial\psi}\right).$$

If U does not depend explicitly on time, the integral of energy exists

$$H = \frac{1}{2}\left(\dot{r}^2 + \frac{L^2}{r^2}\right) + \Pi - U = h - U. \tag{2.78}$$

2.4.3 Relative motion of a tethered system

As the equation of motion of the tethered system about the centre of mass (1.4) and the equation of motion of its centre of mass (1.5) coincide with the form of the equation of motion of a material point on the elastic connection under the action of perturbations, each of these motions can be described by the equations of perturbed motion of type (2.68) – (2.76).

As it is known, the equations of perturbed motion of the centre of mass, by virtue of the periodicity of motion and presence of the obvious dependence between linear and angular variables in unperturbed motion can be transformed into a simpler form. The equation of perturbed relative motion in some modes of motion can be also transformed into a simpler form.

Let us consider the practically important case of motion of a tethered system with small amplitude of longitudinal oscillations in the mode of motion without losing connection. In this case the longitudinal oscillations of the tether are close to harmonic ones, and the equations describing longitudinal oscillations are obtained in a simpler form. In fact, if $\{(r - r_0)/r_0\}^2 \ll 1$, where r_0 is defined from the equality of accelerations following from elastic and centrifugal forces (2.25), then by relating the terms of order $\{(r - r_0)/r_0\}^2$ in the decomposition of the centrifugal acceleration to perturbing accelerations we obtain that in unperturbed motion r performs harmonic oscillations near the value r_0 (2.31). Taking into account that for harmonic oscillations $h = \frac{1}{2}k^2b^2$, where k and b are the frequency and the amplitude of oscillations, according to the technique of derivation of the equations of perturbed motion we obtain

$$\dot{b} = \frac{\dot{h}}{k^2 b} = -\frac{\sin \omega (F_1 + F_{cm})}{k} +$$

$$\left[\frac{b}{r_0} \sin^2 \omega \left(-3 + 12 \frac{L^2}{r_0^4} k^2 \right) + 2 \cos \omega \right] \frac{L\dot{L}}{r_0^3 k^2}, \qquad (2.79)$$

$$\dot{\omega} = k + (\dot{r}_0 + b \cos \omega)/b \sin \omega,$$

where

$$\dot{r}_0 = \frac{2L\dot{L}}{r_0^3 k^2}, \quad k = \sqrt{\frac{c_m}{d} + 3 \frac{L^2}{r_0^4}},$$

F_{cm} is defined by the formula (2.32).

The earlier derived equations of perturbed motion are correct in the non-rotating system of coordinates. However, it is frequently convenient to consider the relative motion in the system of coordinates connected to the orbital motion, because in such a case expressions of projections of the perturbing forces become significantly simpler, and the interpretation of results of the analysis of the equations becomes simpler too. The systems of coordinates connected to the orbital motion are rotating in the general case and consequently the equations of perturbed motion relative to these coordinates are slightly changed.

2.4.4 Motion about the orbit of the mass centre

Let us introduce the right-handed systems of coordinates: $C\xi\eta\zeta$ is the motionless system of coordinates with its origin in the attracting centre C. $CXYZ$ is a "perigee" system of coordinates, connected to the instant orbit of motion of the mass centre, the axis CX is directed from the attracting centre to the pericentre of the instant orbit, the axis CY lies in the plane of the instant orbit and is directed to the centre of mass of the motion in the pericentre, the axis CZ is in the direction of the vector of the moment of momentum of the orbital motion (Fig. 2.12). $Oxyz$ is the moving system of coordinates with origin in the mass centre O of the tethered system. The axis Oz has the direction of the vector of moment of momentum of the relative motion, axis Ox the direction of the vector $\vec{r}$.

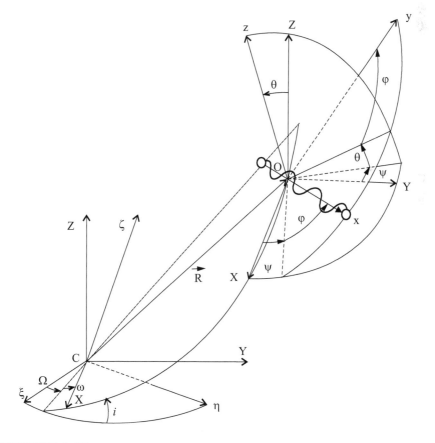

FIGURE 2.12
Coordinate systems.

The mutual orientation of the systems $C\xi\eta\zeta$ and $CXYZ$, $CXYZ$ and $Oxyz$ is defined by Eulerian angles Ω, i, ω_π (the longitude of ascending node, the inclination of the orbit, the argument of the pericentre respectively) and ψ, θ, φ.

Following the general technique we obtain that the equations of perturbed motion of the system about the evolving orbit are equations (2.48) and (2.50), where ω_{01}^*, ω_{02}^*, ω_{03}^* are projections of the angular-velocity vector of the "perigee" coordinate system in the inertial system on to the axes of the first one:

$$\omega_{01}^* = \dot{\Omega} \sin i \sin \omega_\pi + \frac{di}{dt} \cos \omega_\pi,$$

$$\omega_{02}^* = \dot{\Omega} \sin i \cos \omega_\pi - \frac{di}{dt} \sin \omega_\pi, \qquad (2.80)$$

$$\omega_{03}^* = \dot{\omega}_\pi + \dot{\Omega} \cos i.$$

The form of the equation describing the change of the distance r (2.69) obviously remains unchanged and accordingly the equations of perturbed longitudinal oscillations (2.70)–(2.76), (2.79) do not change.

Let the system of coordinates $CX^*Y^*Z^*$ be considered as the orbital system of coordinates connected to orbital motion: the axis CX^* is in the direction of the instant radius-vector of the orbit, the axis CY^* is directed parallel to the transversal of the orbit, the axis CZ^* in the direction of the vector of moment of momentum of the orbital motion. Then the form of equations (2.78) does not change and only the values of projections of $\vec{\omega}^*$ change:

$$\omega_{01}^* = \dot{\Omega} \sin i \sin u_0 + \frac{di}{dt} \cos u_0,$$

$$\omega_{02}^* = \dot{\Omega} \sin i \cos u_0 - \frac{di}{dt} \sin u_0, \qquad (2.81)$$

$$\omega_{03}^* = \dot{u}_0 + \dot{\Omega} \cos i.$$

where $u_0 = \omega_\pi + \nu$ and ν is the argument of latitude and true anomaly of the orbital motion.

Thus, equations of perturbed motion of the tethered system about the evolving orbit are constructed. These equations and the equations of the perturbed motion of the mass centre form the complete system of equations of perturbed motion of the tether, the system of equations allowing one to carry out research making use of the method of averaging.

2.4.5 Motion of the mass centre

The equations of perturbed motion of the mass centre (the perturbed Keplerian motions) are known. However, the used algorithm of derivation of the equations of perturbed Keplerian motion differs from the scheme offered above. We show [93] that the use of the offered scheme for the derivation of the equations of perturbed Keplerian motion allows to reduce the number of necessary transformations in comparison with derivation of similar equations in [45, 70].

Taking into account that in Keplerian motion $L = \sqrt{\mu p}$, p is the focal parameter of the orbit and changing the designation of Eulerian angles from (2.68) we obtain

$$\frac{di}{dt} = \frac{R}{p}\cos u \widetilde{F}_{03}, \quad \dot{\Omega} = \frac{R \sin u}{p \sin i} \widetilde{F}_{03},$$

$$\dot{p} = 2R\widetilde{F}_{02}, \quad \dot{u} = \frac{\sqrt{\mu p}}{R^2} - \frac{R}{p} \sin u \cot i \widetilde{F}_{03}, \tag{2.82}$$

where

$$\widetilde{F}_{03} = \sqrt{\frac{p}{\mu}} F_{03}, \widetilde{F}_{02} = \sqrt{\frac{p}{\mu}} F_{02},$$

F_{03}, F_{02} are accordingly normal and transversal perturbing accelerations of the mass centre.

It is natural to represent relations (2.61) in Keplerian motion as

$$\begin{aligned} R &= \frac{p}{1 + e \cos \nu}, & \frac{d\nu}{dt} &= \frac{\sqrt{\mu p}}{R^2}, \\ \dot{R} &= \sqrt{\frac{\mu}{p}} e \sin \nu, \end{aligned} \tag{2.83}$$

where e is the eccentricity of the orbit.

Taking into account that

$$h = \frac{\mu}{2a} = \frac{\mu(1 - e^2)}{2p},$$

from (2.70) it is easy to obtain

$$\dot{a} = \frac{2a^2 e \sin \nu}{p} \widetilde{F}_{01} + \frac{2a^2 \widetilde{F}_{02}}{R},$$

$$\dot{e} = \widetilde{F}_{01} \sin \nu + \left[\cos \nu + (e + \cos \nu)\frac{R}{p}\right] \widetilde{F}_{02}, \tag{2.84}$$

$$\dot{\nu} = \frac{\sqrt{\mu p}}{R^2} + \frac{\widetilde{F}_{01} \cos \nu}{e} - \widetilde{F}_{02}\left(1 + \frac{R}{p}\right)\frac{\sin \nu}{e},$$

where

$$\widetilde{F}_{01} = \sqrt{\frac{p}{\mu}} F_{01},$$

F_{01} is the radial perturbing acceleration of the mass centre.

From (2.82), (2.84) it follows

$$\dot{\omega}_\pi = \dot{u} - \dot{\nu} = -\frac{\widetilde{F}_{01} \cos \nu}{e} + \widetilde{F}_{02}\left(1 + \frac{R}{p}\right)\frac{\sin \nu}{e} - \frac{r}{p}\sin u \cot i \widetilde{F}_{03}. \tag{2.85}$$

The equation for the time to pass through the perigee τ is defined by differentiation of the equality obtained from (2.83) with respect to time

$$t - \tau = \frac{p^{3/2}}{\sqrt{\mu}} \int\limits_{0}^{\nu} \frac{d\nu}{(1 + e \cos \nu)^2}. \tag{2.86}$$

Below, according to the purpose of research we use the equations of perturbed Keplerian motion in the following form

$$\frac{di}{dt} = \frac{R}{p} \cos u \, \widetilde{F}_{03}, \quad \dot{\Omega} = \frac{R \sin u}{p \sin i} \widetilde{F}_{03},$$

$$\dot{p} = 2R\widetilde{F}_{02}$$

$$\dot{e} = \widetilde{F}_{01} \sin \nu + \left[\cos \nu + (e + \cos \nu)\frac{R}{p}\right] \widetilde{F}_{02}, \tag{2.87}$$

$$\dot{\omega}_\pi = -\frac{\widetilde{F}_{01} \cos \nu}{e} + \widetilde{F}_{02}\left(1 + \frac{R}{p}\right)\frac{\sin \nu}{e} - \frac{R}{p}\sin u \cot i \, \widetilde{F}_{03},$$

$$\dot{u} = \frac{\sqrt{\mu p}}{R^2} - \frac{R}{p}\sin u \cot i \, \widetilde{F}_{03}.$$

2.5   On the derivation of new forms of equations of perturbed Keplerian motion

In the previous section ordinary differential equations of perturbed Keplerian motion were derived, in the form of Newton's equations [46]. The scheme of their derivation is based on the use of the theorem of change of moment of momentum and on the technique of construction of the equations of perturbed motion of the non-linear oscillatory element [90]. The main distinction of the offered technique consists in the fact that in the derivation of the equations of perturbed motion of radial oscillations not the first integrals of motion, but those oscillations which are most suitable for the investigation of the particular task, are used as the basic function.

In [70] as basis of the derivation of the equations of perturbed Keplerian motion, not first integrals but formulas directly describing the change of variables in unperturbed motion were used. It has allowed to considerably simplify the derivation of the equations of perturbed Keplerian motion in comparison with earlier known forms. However, awkward differentiation of vectorial values carried out in [70], as we see, is not necessary.

Celestial mechanics, developed during centuries by outstanding scientists, contains a large variety of the equations of perturbed Keplerian motion [46]. These equations comprise practically all cases of orbital motion of satellites, and they are used in the majority of problems of mechanics of space flight. However, the mechanics of the space flight, especially nowadays, puts problems, which have already essential differences from the classical problems of celestial mechanics. Let us show on relatively simple examples that the developed technique allows to build the new forms of the equations of perturbed Keplerian motion. At the same time, for some problems of mechanics of space flight, the formulation of which is different from the tasks of the celestial mechanics, the classical forms of the equations are represented too much complex, and it can appear effective to introduce new variables describing the orbital motion. At the same time, earlier known schemes for the derivation of equations of perturbed Keplerian motion, such as the scheme based on the evaluation of brackets of Lagrange for elliptical orbit elements, introduced in the course of celestial mechanics by Tisserand [112], Duboshin's scheme [45], based on geometrical constructions, and Lurie's scheme [70], based on the fixation of vectorial expressions of unperturbed motion, are represented excessively bulky which hampers the introduction of new variables, convenient for particular problems.

Let us show by examples [93] that the developed technique allows rather simply to build new forms of the equations of perturbed Keplerian motion. Let us consider the motion of the satellite on a near-circular orbit. The deduced equations of perturbed motion have a singularity at $e = 0$, therefore their use in the case of eccentricities, close to zero is inconvenient. Besides, the concept of eccentricity for orbits of satellites around the Earth close to a circle often has not the special sense, since the deviations taken into account by the eccentricity of the orbit from the circle are commensurable with the deviation of orbits from Keplerian ones, introduced by disturbances, for example, by the eccentricity of the gravitational field of the Earth.

Classically, in case of small eccentricities, in the equations obtained for the general case, new variable are introduced, the variables of Lagrange λ_1 and λ_2, $\lambda_1 = e \sin \omega_\pi$, $\lambda_2 = e \cos \omega_\pi$. Thus the equations of perturbed motion do not have singularities at $e = 0$, but as reference trajectory for an almost circular orbit, Keplerian motion on an elliptical orbit is chosen.

Since motions on elliptical and circular orbits have qualitative differences, the introduction of variables of Lagrange results in complication of the equations. For introduction of variables of Lagrange it is necessary to pay for their redundant awkwardness. Thus, for example, λ_1, λ_2, describing the form of the orbit in these equations, depend on the normal component of the perturbing forces, though it is obvious that the normal component should not influence the form of the orbit. Certainly, the techniques of research of these equations developed in celestial mechanics and the obtained results quite pay back for the redundant awkwardness of these equations.

From the point of view of mechanics in the case of an almost circular orbit of motion of a satellite as reference orbit of unperturbed motion it is natural to consider a circular Keplerian orbit.

In this case instead of (2.83) let us suppose

$$R = p\,(1 + b_1), \quad \dot{R} = b_2\sqrt{\mu/p}, \qquad (2.88)$$

and consider (2.88) as the formulas of introduction of the new variables b_1, b_2. By virtue of the statement of the problem b_1 and b_2 are small, $b_i \ll 1, (i = 1, 2)$.

Differentiating the first equation (2.88) with respect to time we obtain

$$\dot{R} = \dot{p}\,(1 + b_1) + p\,\dot{b}_1, \ \Rightarrow \ \dot{b}_1 = b_2\sqrt{\mu/p^3} - 2(1 + b_1)^2 F_{02}.$$

Differentiating the second equality (2.88), taking into account (2.69), (2.82), we obtain

$$\ddot{R} = \dot{b}_2\sqrt{\frac{\mu}{p}} - \frac{1}{2}b_2\sqrt{\frac{\mu}{p^3}}\dot{p} = \frac{\mu p}{R^3} - \frac{\mu}{R^2} + \tilde{F}_{01}, \ \Rightarrow$$

$$\dot{b}_2 = b_2(1 + b_1)\tilde{F}_{02} + \tilde{F}_{01} - \sqrt{\frac{\mu}{p^3}}\frac{b_1}{(1 + b_1)^3}.$$

Thus, for an orbit close to a circular one the equations of perturbed Keplerian motion look like

$$\dot{\Omega} = z\frac{\sin u}{\sin i}\tilde{F}_{03}, \quad \frac{di}{dt} = z\cos u\tilde{F}_{03},$$

$$\dot{\gamma} = 2zs\tilde{F}_{02},$$

$$\Delta\dot{u} = \sqrt{\frac{\mu}{p_0^3}}\left(\frac{1}{s^{3/2}z^2} - 1\right) - \dot{\Omega}\cos i, \qquad (2.89)$$

$$\dot{b}_1 = b_2\sqrt{\frac{\mu}{p^3}} - 2z^2\tilde{F}_{02},$$

$$\dot{b}_2 = b_2z\tilde{F}_{02} + \tilde{F}_{01} - \sqrt{\frac{\mu}{p^3}}\frac{b_1}{z^3},$$

where $z = 1 + b_1$, $s = 1 + \gamma$, $p = p_0(1 + \gamma)$, $\Delta u = u - u_0$, p_0, u_0 are accordingly focal parameter and the argument of latitude of the referenced unperturbed circular orbit, $\dot{u}_0 = \sqrt{\mu/p_0^3}$, γ is small value following from the problem statement.

After the introduction of the transformation of variables

$$R = p_0(1 + b_1), \quad \dot{R} = b_2\sqrt{\frac{\mu}{p_0}} \qquad (2.90)$$

the equations of motion take the form

$$\dot{\Omega} = zs^{-1}\widetilde{F}_{03}\frac{\sin u}{\sin i},$$

$$\frac{di}{dt} = zs^{-1}\widetilde{F}_{03}\cos u,$$

$$\Delta\dot{u} = \sqrt{\frac{\mu}{p_0^3}}\left(\frac{\sqrt{s}}{z^2}-1\right)-\dot{\Omega}\cos i,$$

$$\dot{b}_1 = b_2\sqrt{\frac{\mu}{p_0^3}},\tag{2.91}$$

$$\dot{b}_2 = \sqrt{\frac{\mu}{p_0^3}}\frac{\gamma-b_1}{z^3}+\sqrt{\frac{p_0}{\mu}}F_{01},$$

$$\dot{\gamma} = 2zs\widetilde{F}_{02},$$

where the same notations, as in (2.89), are used.

The obtained new forms of equations (2.89), (2.91) can effectively be used in the task of fast numerical analysis of motion of a satellite on orbits close to a circle, when the results obtained in celestial mechanics can not be used (for example, in the problems of development both of the analysis of attitude control systems and stabilisation of satellites about the mass centre, when the calculation of parameters of orbital motion is necessary; usually this calculation does not require high accuracy, but often, for example, because of action of aerodynamic forces, it is impossible to take advantage of the approximated final formulas).

Let us point out that it is better to execute the numerical integration of equations (2.89), (2.91) with respect to the new independent variable u_0, the argument of latitude of the unperturbed orbit.

Let us also point out that equations (2.89), (2.91) allow essentially simplify obtaining the final approximated formulas of change of orbit parameters. Hence, the known formulas of disturbances of orbital motion from the second zonal harmonics [46] for circular orbits for equations (2.91) are obtained simply as the result of the first iteration of an allocation procedure.

Let us consider now the problem of motion of two bodies on the neighboring orbits. This problem is relevant for understanding of motions of groups of satellites and the motion of a basic satellite and a subsatellite. In these problems it is natural to describe the motion of other bodies in relation to the orbit of the basic satellite (to the basic orbit). As the inclination of orbits is small, the use of the Euler angles is unacceptable. In celestial mechanics, in this case artificial reception (introduction of an average longitude or Lagrange's variables) for the avoidance of degeneration in the equations is used [46].

From the point of view of mechanics such a method is considered a little bit artificial. Everything said before about the introduction of variables of Lagrange for the motion of satellites on almost circular orbits is valid also in this case. For small inclinations it is natural to use another triple of angles of orientation, namely angles convenient for small inclinations of the basic planes.

Let us introduce the right-handed coordinate systems with the origin in the attracting centre C. $CX^*Y^*Z^*$ is a non-rotating coordinate system. $CXYZ$ is the coordinate system connected to the basic orbit, and the axis CZ is directed to the binormal of the orbit. $Cxyz$ is the coordinate system connected to the motion of the subsatellite. The axis Cx is directed from the attracting centre to the mass centre of the subsatellite, the axis Cz is in the direction of the vector of moment of momentum of its orbital motion (binormal of the orbit).

Let us select Bryant angles φ_1, φ_2, φ_3 [126] (aircraft angles, φ_1, φ_2, φ_3 — respectively yaw, pitch and roll angles) as angles of orientation of $Cxyz$ in $CXYZ$.

Let $\vec{\omega}$ be the angular-velocity vector $Oxyz$ in $OXYZ$, and $\vec{\omega}^*$ be the angular-velocity vector $OXYZ$ with respect to $OX^*Y^*Z^*$. Then on the basis of the theorem of change of moment of momentum we obtain

$$\vec{L}' + (\vec{\omega} + \vec{\omega}^*) \times \vec{L} = \vec{R} \times \vec{F}^*. \tag{2.92}$$

Using the relations between projections of the angular velocity on the axis of the moving system of coordinates and derivatives of the Bryant's angles

$$\omega_1 = \dot{\varphi}_1 \cos\varphi_2 \cos\varphi_3 + \dot{\varphi}_2 \sin\varphi_3,$$

$$\omega_2 = -\dot{\varphi}_1 \cos\varphi_2 \sin\varphi_3 + \dot{\varphi}_2 \cos\varphi_3,$$

$$\omega_3 = \dot{\varphi}_1 \sin\varphi_2 + \dot{\varphi}_3,$$

we obtain similarly to (2.48), (2.50)

$$\dot{\varphi}_1 = \frac{RF_{03}}{L} \frac{\cos\varphi_3}{\cos\varphi_2} + \frac{1}{\cos\varphi_2} (\omega_2^* \sin\varphi_3 - \omega_1^* \cos\varphi_3),$$

$$\dot{\varphi}_2 = \frac{RF_{03}}{L} \sin\varphi_3 - (\omega_1^* \sin\varphi_3 + \omega_2^* \cos\varphi_3), \tag{2.93}$$

$$\dot{\varphi}_3 = \frac{L}{R^2} - \omega_3^* - \dot{\varphi}_1 \sin\varphi_2.$$

Or expressing ω_1^*, ω_2^*, ω_3^* through projections $\vec{\omega}^*$ on axes $CXYZ$ ω_{01}^*, ω_{02}^*, ω_{01}^*, we obtain

$$\dot{\varphi}_1 = \frac{R \cos\varphi_3}{p \cos\varphi_2} \widetilde{F}_{03} - \omega_{01}^* - \tan\varphi_2(\omega_{02}^* \sin\varphi_1 - \omega_{03}^* \cos\varphi_1),$$

$$\dot{\varphi}_2 = \frac{R}{p} \sin\varphi_3 \widetilde{F}_{03} - \omega_{02}^* \cos\varphi_2 - \omega_{03}^* \sin\varphi_1,$$

$$\dot{\varphi}_3 = \frac{\sqrt{\mu p}}{R^2} + \frac{1}{\cos\varphi_2} [\omega_{02}^* \sin\varphi_1 - \omega_{03}^* \cos\varphi_1 -$$

$$\frac{R}{p} \cos\varphi_3 \sin\varphi_2 \widetilde{F}_{03}]. \tag{2.94}$$

Depending on the system of coordinates connected to the orbital motion of the basic satellite, ω_{01}^*, ω_{02}^*, ω_{03}^* have different forms. In the orbital system of coordinates

$$
\begin{aligned}
\omega_{01}^* &= \dot{\Omega}_a \sin i_a \sin u_a + \frac{di_a}{dt} \cos u_a, \\
\omega_{02}^* &= \dot{\Omega}_a \sin i_a \cos u_a - \frac{di_a}{dt} \sin u_a, \\
\omega_{03}^* &= \dot{u}_a + \dot{\Omega}_a \cos i_a.
\end{aligned}
\tag{2.95}
$$

In the "perigee" system of coordinates [16] the axis CX is directed along the position vector of the pericentre of the basic orbit. We have

$$
\begin{aligned}
\omega_{01}^* &= \dot{\Omega}_a \sin i_a \sin \omega_{\pi a} + \frac{di_a}{dt} \cos \omega_{\pi a}, \\
\omega_{02}^* &= \dot{\Omega}_a \sin i_a \cos \omega_{\pi a} - \frac{di_a}{dt} \sin \omega_{\pi a}, \\
\omega_{03}^* &= \dot{\omega}_{\pi a} + \dot{\Omega}_a \cos i_a.
\end{aligned}
\tag{2.96}
$$

In the half-connected system $CX^*Y^*Z^*$, for which axis CX^* lies in the equatorial plane of the Earth and coincides with the direction of the ascending node of the basic orbit,

$$
\omega_{01}^* = \frac{di_a}{dt}, \qquad \omega_{02}^* = \dot{\Omega}_a \sin i_a, \qquad \omega_{03}^* = \dot{\Omega}_a \cos i_a.
\tag{2.97}
$$

The index "a" designates the variables appropriate to the motion on the basic orbit (of the basic satellite). In all cases following from the formulation of the task φ_1, φ_2 are small values.

In order to obtain the complete equations of perturbed relative motion of the subsatellite, the equation (2.92) should be supplemented by the equations of relative change of the focal parameter and the distance from the attracting centre.

We enter new variables δ, b_{C1}, b_{C2} as follows

$$
p_c = p_a(1 + \gamma_c), \quad R_c = R_a(1 + b_{c1}), \quad \dot{R}_c = \dot{R}_a + b_{c2}\sqrt{\frac{\mu}{R_a}}.
\tag{2.98}
$$

The index "c" designates variables of orbital motion of the subsatellite.

Because of the explained technique it is possible to obtain the following equations:

$$\dot{\gamma}_c = 2\frac{R_a}{p_a}\left[(1+b_{c1})\widetilde{F}_{02c} - \widetilde{F}_{02a}(1+\gamma_c)\right],$$

$$\dot{b}_{c1} = b_{c2}\sqrt{\frac{\mu}{R_a^3}} - \frac{\dot{R}_a}{R_a}b_{c1},$$

$$\dot{b}_{c2} = \sqrt{\frac{\mu}{R_a^3}}\left[\frac{p_a}{R_a}\left(\frac{1+\gamma_c}{(1+b_{c1})^3}-1\right)+1-\frac{1}{(1+b_{c1})^2}\right] +$$

$$\sqrt{\frac{R_a}{\mu}}(F_{01c}-F_{01a}) + \frac{1}{2}b_{c2}\frac{\dot{R}_a}{R_a}, \tag{2.99}$$

Equations (2.94), (2.96) are equations of perturbed motion of the subsatellite varying about the orbit of the basic satellite. They may be easily linearized along the basic orbit.

For the motion on a circular orbit it is possible to use the following transformation of variables

$$p_c = p_a(1+\gamma_c), \quad R_c = p_a(1+b_{c1}), \quad \dot{R}_c = b_{c2}\sqrt{\frac{\mu}{p_a}}. \tag{2.100}$$

Then

$$\dot{\gamma}_c = 2(1+b_{c1})\widetilde{F}_{02c} - 2(1+b_{a1})\widetilde{F}_{02a}(1+\gamma_c),$$

$$\dot{b}_{c1} = b_{c2}\sqrt{\mu/p_a^3} - 2(1+b_{c1})(1+b_{a1})\widetilde{F}_{02a}, \tag{2.101}$$

$$\dot{b}_{c2} = b_{c2}(1+b_{a1})\widetilde{F}_{02a} + \sqrt{\frac{\mu}{p_a^3}}\frac{\gamma_c-b_{c1}}{(1+b_{c1})^3} + \sqrt{\frac{p_a}{\mu}}F_{01c},$$

where $\gamma_c \ll 1$ and $b_i \ll 1$ by virtue of the statement of the tasks. Equations (2.94), (2.101) allow with the use of (2.91) to construct the effective computational scheme.

3

Analysis of the Motion of TSS

3.1 Regular attitude motions of TSS

3.1.1 On application of the averaging method

The method of averaging is a mathematically justified method of research and it is based on a number of strictly proven theorems [32, 51, 80, 123]. On the other hand, the method of averaging is based on the wide evidence of its successful practical application in the investigation of various problems. And due to its "physicality" the method of averaging was used in celestial mechanics and for problems of non-linear oscillations much earlier than it has been mathematically justified. Now the domain of application of the method of averaging is a little wider than the domain where it is justified. Thus, generally speaking, the results of researches can be incorrect and hence direct comparison with results of numerical integration of the complete equations is the only way of checking its validity.

Now the method of averaging is a ramified, well advanced tool of applied research containing various algorithms (schemes, operators) of averaging. Simultaneously the formal application of these algorithms is possible only in cases of rather simple systems of equations (for example, with one rotating phase [82]). For the investigation of rather complex systems the application of the method of averaging assumes the determination of the area of the investigated motion, the mode of motion and the choice of the most suitable algorithm of averaging. In other words, the application of the method of averaging for the investigation of rather complex systems represents a certain problem. The attempts of formal application of the method of averaging sometimes does not give meaningful results [77].

Thus, for the research of complex problems of dynamics by the method of averaging it is necessary to correlate the system of the equations of the problem with mathematically justified systems, to carry out the choice of the scheme of averaging and to reduce the system of the equations of the problem to the appropriate form. As the formulation of the theorems of the method of averaging and their proof is given in the language of mathematical analysis, it is necessary to estimate numerically real deviations of averaged trajectories from true ones. The successful application of the method of averaging is possible also when the problem is not resulting into mathematically reasonable cases. Thus the qualitative analysis of possible modes of motion and the strict numerical control of results are especially necessary.

The right parts of the equations of perturbed motion of the considered system, generally speaking, are periodic with respect to each of the three angular variables: ω is the phase of longitudinal oscillations, φ is the angle of pure rotation of the system about the mass centre, ν is the angle of pure rotation of the mass centre. I.e., the equations of perturbed motion contain three rotating phases. The relations between average frequencies of oscillations of the system on these variables depend on initial parameters of motion and elastic properties of the tether and can be various. Therefore methods of research of perturbed motion should be various also.

We from now on assume that the stiffness of the tether is high enough and the average frequency of longitudinal oscillations significantly surpasses the frequency of the average orbital motion. Concerning the average frequency of rotation of the tethered system around the mass centre depending on the considered mode of motion various assumptions are made. Basic attention is given to the evolution of parameters of motion of the system quickly rotating about the mass centre, i.e., to the mode of its rotational motion about the mass centre if the average frequency of rotation of the tethered system significantly surpasses the frequency of average orbital motion.

In such a statement of the problem the equations of motion of the system refer to the most difficult class to be researched, the so-called "strongly perturbed systems" [51], the methods of research of which are far from being complete. A classical example of a strongly perturbed system is the problem of the motion of the mass centre of the Moon in the field of attraction of the Earth and the Sun. The history of construction of satisfactorily accurate solutions of this problem comprises more than a hundred years.

It is known [51] that the application of the method of averaging to strongly perturbed systems depending on the operator of smoothing and the choice of generating (nonzero) solutions can give both good and incorrect (even absurd) results. Therefore, depending on the relation of average frequencies and the mode of motion of the system we apply various operators of smoothing: the operator of averaging along the generating solution and the operator of averaging with respect to the fast variable. We compare the solution of the averaged equations with results of numerical solutions of the complete equations of motion in each case.

3.1.2 Influence of gravitational oscillations

It is everywhere supposed in this chapter that the values of order $\varepsilon_1{}^2$ are negligible small, where $\varepsilon_1 = r/R$ (Fig. 1.3). The influences of the central Newtonian field of forces on the relative motion of the tethered system with accuracy to the members up to $\varepsilon_1{}^2$ are described by the perturbed function

$$U = \frac{1}{2}\frac{\mu}{R^3}r^2\left(3(\vec{e_r}, \vec{e_R})^2 - 1\right), \tag{3.1}$$

where $\vec{e}_r = \vec{r}/r$, $\vec{e}_1$ is the unit vector of the axis Ox, $\vec{e}_R = \vec{R}/R$ is the unit vector of the local vertical. The trajectory of the mass centre with the same accuracy is the unperturbed Keplerian orbit.

The projections of the perturbed acceleration of the Newtonian field of forces on the axes of the moving system of coordinates are

$$F_1 = \frac{\mu}{R^3} r \left(3(\vec{e_1}, \vec{e_R})^2 - 1 \right),$$

$$F_2 = 3\frac{\mu}{R^3} r (\vec{e_1}, \vec{e_R})(\vec{e_2}, \vec{e_R}), \qquad (3.2)$$

$$F_3 = 3\frac{\mu}{R^3} r (\vec{e_1}, \vec{e_R})(\vec{e_3}, \vec{e_R}),$$

where $\vec{e_i}$ are unit vectors of the moving system of coordinates, $(\vec{e}_1, \vec{e}_R) = \cos\theta \sin\varphi \sin(\nu - \psi) + \cos\varphi \cos(\nu - \psi), (\vec{e}_3, \vec{e}_R) = -\sin\theta \sin(\nu - \psi), (\vec{e}_2, \vec{e}_R) = \cos\theta \cos\varphi \cos(\nu - \psi) \sin\varphi \cos(\nu - \psi)$. One can obtain equation (3.2) either directly from equation (2.2), or from equation (3.1) with help of equation (2.77).

We assume that the angular velocity of rotation of the system about its mass centre is essentially larger than the angular velocity of its orbital motion, namely, the square of the ratio of the second one to the first one is a small value $\dfrac{\mu}{p^3} \dfrac{r^4}{L^2} = \varepsilon_2 \ll 1$. Concerning the velocity of change of the phase of longitudinal oscillations we assume that its value has the order not smaller than the value of angular velocity of relative rotation.

3.1.3 Motion due to longitudinal oscillations of small amplitude

We consider in the beginning the mode of motion without loss of tension (without disappearance of the stress of stretching of the tether) with, in comparison with length of the tether, small amplitude of longitudinal oscillations of the system $(b/r_0)^2 = \varepsilon_3 \ll 1$ [5]. Such a mode of motion is supposed in most projects of use of TSS, and simultaneously it is the simplest for research. The equations of motion in the considered mode contain three small values ε_1, ε_2 and ε_3, and the equations of perturbed motion have a simpler form (2.68), (2.79). The generating solution is represented by the formulas (2.31), (2.34).

We average the equations of perturbed motion along the generating solution keeping only terms of first order of smallness with respect to the introduced small values. Thus, in the studied mode of motion according to the analysis of unperturbed motion the resonances of lowest orders 1:1 and 1:2 cannot take place. Therefore for any ratio of the average frequencies of the longitudinal oscillations and the relative rotation of the system the equations of first approximation are correct:

$$\dot{\psi} = N_1 \cos\theta \sin^2(\nu - \psi),$$

$$\dot{\theta} = N_1 \sin\theta \cos(\nu - \psi) \sin(\nu - \psi),$$

$$\dot{\nu} = \sqrt{\frac{\mu}{p^3}}(1 + e\cos\nu)^2, \tag{3.3}$$

$$\dot{L} = 0, \quad \dot{b} = 0,$$

where

$$N_1 = -\frac{3}{2}\frac{\mu}{p^3}\frac{r_0^2}{L}(1 + e\cos\nu)^3.$$

Therefore, in the considered mode the influence of the Newtonian field of forces does not change the amplitude of the longitudinal oscillations and the value of the moment of momentum of the system in first approximation.

To obtain equation (3.3) one can substitute the generating solution (2.31), (2.34) into the equations of the perturbed motion (2.68), (2.79), and taking into account expressions (3.2) for perturbing forces, apply the averaging operator along the generating solution.

Comparing equation (3.3) with similar equations of first approximation for a symmetric rigid body [16, 18] it is easy to see that they coincide up to notations. This means that for the considered mode of motion of the system the evolution of the moment of momentum of the system and the dumb-bell with length of the bar r_0 coincides in first approximation.

We pass in the system of equations (3.3) from the independent variable t to the new independent variable ν by making use of the dependence

$$d\nu = \sqrt{\frac{\mu}{p^3}}(1 + e\cos\nu)^2 dt,$$

which we obtain from the third equation of (3.3):

$$\frac{d\psi}{d\nu} = N_0(1 + e\cos\nu)\cos\theta\sin^2(\nu - \psi),$$

$$\frac{d\theta}{d\nu} = N_0(1 + e\cos\nu)\sin\theta\cos(\nu - \psi)\sin(\nu - \psi). \tag{3.4}$$

Here $N_0 = -\dfrac{3}{2}\sqrt{\dfrac{\mu}{p^3}}\dfrac{r_0^2}{L}$.

The detailed analysis of these equations is given in [16]. The equations (3.4) can be analytically integrated only in the specific case of a circular orbit of motion of the mass centre. The basic evolutionary effects in the general case are determined by averaging of the equations (3.4) with respect to variable ν

$$\frac{d\theta}{d\nu} = 0, \quad \frac{d\psi}{d\nu} = -\frac{3}{4}\sqrt{\frac{\mu}{p^3}}\frac{r_0^2}{L}\cos\theta. \tag{3.5}$$

Thus, the basic influence of the central Newtonian field of forces on the motion of the system in the considered mode is determined by the fact that "the vector of moment of momentum performs a precession with constant angular velocity around the normal to the plane of orbit, keeping the angle between the two vectors constant" [18].

The accuracy of equations (3.5) is much lower than the accuracy of equations (3.4), since equations (3.5) are the equations of first approximation with respect to the small parameter

$$\sqrt{\frac{\mu}{p^3} \frac{r_0^2}{L}} = \sqrt{\varepsilon_2}.$$

In Figs 3.1–3.3 characteristic plots of the deviations of the solutions of the averaged equations of first (equation (3.3) and second equation (3.5)) averaging of the solutions of the non-averaged equations are shown. The calculations were carried out for the following parameters: parameters of the orbit of motion of the mass centre: $e = 0.2$, $p = 7885$ km; parameters of the system: $d = 1000$ m, $c_m/d = 0.1\,\text{s}^{-2}$; initial values: $L/d^2 = 0.18\,\text{s}^{-1}$, $r = r_1 = 1080$ m, $\dot{r} = 0$, $\psi = \pi/4$, $\theta = \pi/6$, $\varphi = 0$, $\nu = 0$. In these calculations and further on, the value of the gravitational parameter is assumed to be equal to the gravitational parameter of the Earth $\mu = 3.986003 \cdot 10^5\,\text{km}^3\text{s}^{-2}$.

For such determined parameters the amplitude of the longitudinal oscillations of the system in unperturbed motion is equal $b \approx 113.9$ m, and $r_0 \approx 1193.9$ m.

In Fig. 3.1 and Fig. 3.2 deviations of the solution of the averaged equations (3.3) (line "Av") from the solution of the exact equations (line "Nonav") for the initial interval of time are represented. It is visible that for all variables the deviations have periodic character. For angles of orientation of the moment of momentum (Fig. 3.1a) and (Fig. 3.1b) the deviations do not surpass 0.00008 rad, and for the specific moment of momentum (Fig. 3.2a) and for the amplitude of longitudinal oscillations (Fig. 3.2b) the deviations are less than 0.01% and 0.04%, respectively. Calculations performed for a long-time interval show that values of deviations of the solution of the equation (3.3) from the solution of the exact equations are kept in prescribed limits, a little bit decreasing for $\nu \approx n\pi$.

In Fig. 3.3 solutions of equations (3.4) and (3.5) are represented. It is visible that deviation of the solution averaged with respect to ν of the equations (3.5) (the lines "AV") from the solution of the equations (3.4) have periodic character with a period that approximately is equal to the orbital period of the system. The maximum deviations for angles ψ (Fig. 3.3a) and θ (Fig. 3.3b) do not surpass 0.07 rad and 0.0016 rad, respectively.

We point out that for the given mode of motion the application of the operator of averaging with respect to the angular variable φ and ω_1 to equations (2.68), (2.74) and also on the angular variable φ^0 and ω_3 to equations (2.76) has the same result, namely equations (3.3).

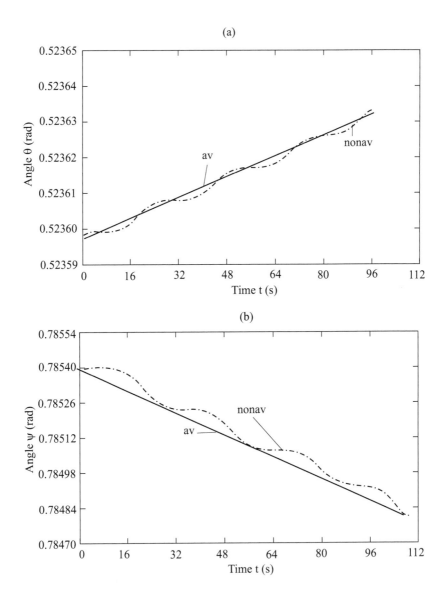

FIGURE 3.1
Variation of motion parameters on an initial interval of time.

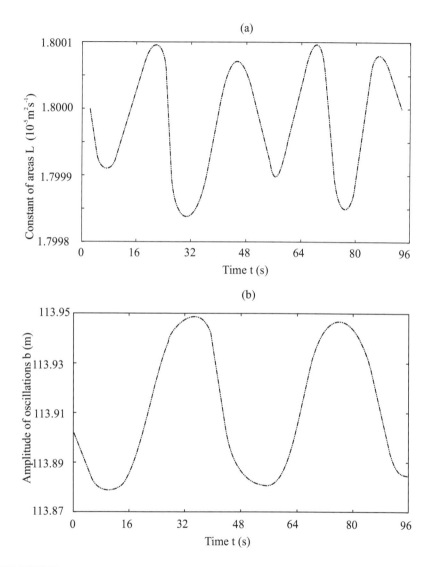

FIGURE 3.2
Variation of motion parameters on an initial interval of time.

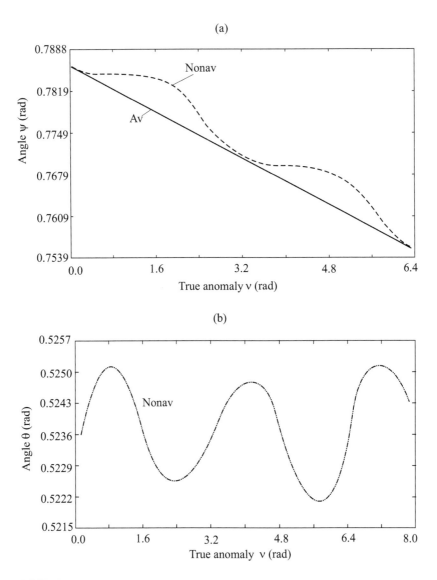

FIGURE 3.3
Deviation of angles ψ and ϑ from their values corresponding to a uniform precession.

3.1.4 Motion with longitudinal oscillations of large amplitude

We consider the mode of motion of the system when the amplitude of the longitudinal oscillations is not small in comparison with the length of the tether and a motion losing connection is possible [87, 90].

For the research of this mode the application of the operator of smoothing along the generating solution is connected to excessively complex transformations and calculations. Thus, even in the case of a rather stiff tether when the amplitude of its stretching is small in comparison with its length, the generating solution can be constructed analytically simply enough, by "pasting" the solutions of free motion and harmonic oscillations for the active tether (2.41), the fulfilment of the operation of averaging along this solution is extremely inconvenient, and obviously it is not possible to carry out this operation analytically.

The most expedient technique of research of the considered mode of motion is the application of the operator of averaging with respect to the angular variable ω_3 and φ^0 to the equations (2.76), since it allows to apply the most developed schemes of averaging, the algorithms of averaging of independent rotating systems.

We consider non-resonance motion of the system with the assumption that ω_{12} and 2π are rationally incommensurable. Then the equations of first approximation obtained by averaging of equations (2.76) with respect to φ^0 and ω_3, look like

$$\frac{d\theta}{d\varphi^0} = N_1^* \sin\theta \cos(\nu - \psi) \sin(\nu - \psi),$$

$$\frac{d\psi}{d\varphi^0} = N_1^* \cos\theta \sin^2(\nu - \psi),$$

$$\frac{d\alpha}{d\varphi^0} = -N_1^* \cos^2\theta \sin^2(\nu - \psi),$$

$$\frac{dL}{d\varphi^0} = 0, \quad \frac{db_v}{d\varphi^0} = 0, \tag{3.6}$$

where

$$N_1^* = -\frac{3}{2}\frac{\mu}{p^3}\frac{r^{*4}}{L^2}(1 + e\cos\nu)^3, \quad r^{*4} = \frac{1}{2\pi}\int_0^{2\pi}\frac{d\omega_3}{(a_v - b_v\Phi_3(\omega_3))^4},$$

and it is assumed that at the initial moment of time $\varphi = \varphi^0$, i.e., $\alpha_0 = 0$. The change of ν does not depend on the relative motion of the system and is determined by equation

$$\frac{d\nu}{dt} = \sqrt{\frac{\mu}{p^3}}(1 + e\cos\nu)^2, \tag{3.7}$$

and the relation between t and φ^0 (and therefore ν and φ^0) is the same as in unperturbed motion

$$\frac{d\varphi^0}{dt} = L\upsilon^2. \tag{3.8}$$

Equations (3.6)–(3.8) represent the complete system of equations of first approximation with respect to small parameters ε_1 and ε_2. From equations (3.6) follows that in first approximation the amplitude of the longitudinal oscillations and the value of the moment of momentum are constant, and the evolution of the orientation of the moment of momentum coincides with the evolution of the orientation of the moment of momentum of a dumb-bell with length of the bar r^* but in "new time" φ^0.

For construction of conformity between t and φ^0 (ν and φ^0) by virtue of the periodicity of ν in φ^0 there is no necessity to integrate equation (3.8) on the whole interval of time. It is enough to know the conformity between t and φ^0 on one period of the longitudinal oscillations:

$$t = \int_{\varphi_0^0}^{\varphi^0} \frac{d\varphi^0}{L\upsilon^2} = \int_{\varphi_0^0}^{\omega_{12}\left[\frac{\varphi^0}{\omega_{12}}\right]} \frac{d\varphi^0}{L\upsilon^2} + \int_{\varphi_0^0}^{\varphi^0 - \omega_{12}\left[\frac{\varphi^0}{\omega_{12}}\right]} \frac{d\varphi^0}{L\upsilon^2} =$$

$$= \frac{\omega_{12}}{2\pi}\left[\frac{\varphi^0}{\omega_{12}}\right]\frac{1}{L}\int_0^{2\pi} \frac{d\omega_3}{(a_\upsilon - b_\upsilon \Phi_3(\omega_3))^2} + \int_{\varphi_0^0}^{\varphi^0 - \omega_{12}\left[\frac{\varphi^0}{\omega_{12}}\right]} \frac{d\varphi^0}{L\upsilon^2} = \tag{3.9}$$

$$= \frac{r_*^2}{L}\omega_{12}\left[\frac{\varphi^0}{\omega_{12}}\right] + \int_{\varphi_0^0}^{\varphi^0 - \omega_{12}\left[\frac{\varphi^0}{\omega_{12}}\right]} \frac{d\varphi^0}{L\upsilon^2},$$

where $[\cdot]$ means the integer part of a number, φ_0^0 is the value of φ in the instant of time $t = 0$.

From (3.9) follows that "time" φ^0 in the instant when it is equal to an integer number of periods of longitudinal oscillations ω_{12} coincides with "time" $\varphi_{av}^0 = Lt/r_*^2$. Therefore ν, determined by equation

$$\frac{d\nu}{d\varphi^0} = \sqrt{\frac{\mu}{p^3}}(1 + e\cos\nu)^2 \frac{r_*^2}{L}, \tag{3.10}$$

differs from the true value by a periodic term with period ω_{12}, the value of which has the order $\sqrt{\varepsilon_2}$. Therefore, the complete system of equations (3.6), (3.10) keep the order of accuracy of ε_2 in relation to the parameters of motion of the system.

Similar reasoning results in the conclusion that the solution of the system of equations

$$\frac{d\theta}{d\varphi_{av}^0} = \frac{r_*^2}{L}\frac{d\theta}{dt} = N_1^* \sin\theta \cos(\nu - \psi)\sin(\nu - \psi),$$

$$\frac{d\psi}{d\varphi_{av}^0} = \frac{r_*^2}{L}\frac{d\psi}{dt} = N_1^* \cos\theta \sin^2(\nu - \psi),$$

$$\frac{d\alpha}{d\varphi_{av}^0} = \frac{r_*^2}{L}\frac{d\alpha}{dt} = -N_1^* \cos^2\theta \sin^2(\nu - \psi),$$

(3.11)

$$\frac{d\nu}{d\varphi_{av}^0} = \frac{r_*^2}{L}\frac{d\nu}{dt} = \sqrt{\frac{\mu}{p^3}}(e + \cos\nu)^2 \frac{r_*^2}{L}$$

differs from the solution of system (3.6)–(3.8) for θ, ψ and α by periodic terms with period ω_{12}, the values of which have the order of ε_2.

Performing in equations (3.6), (3.10) or in equations (3.11) differentiation with respect to ν we obtain equations similar to (3.4)

$$\frac{d\psi}{d\nu} = N_0^*(1 + e\cos\nu)\cos\theta \sin^2(\nu - \psi),$$

$$\frac{d\theta}{d\nu} = N_0^*(1 + e\cos\nu)\sin\theta \cos(\nu - \psi)\sin(\nu - \psi),$$

(3.12)

$$\frac{d\alpha}{d\nu} = -N_0^*(1 + e\cos\nu)\cos^2\theta \sin^2(\nu - \psi),$$

$$N_0^* = -\frac{3}{2}\sqrt{\frac{\mu}{p^3}}\frac{r^{*4}}{r_*^2}\frac{1}{L},$$

the solution of which differs from the solutions of the corresponding equations of first approximation by a periodic term, the value of which has the order of ε_2.

Therefore, the evolution of the orientation of the moment of momentum of the system in first approximation may be determined as the evolution of the moment of momentum of a dumb-bell with the length of the bar equal to $\sqrt{r^{*4}/r_*^2}$.

The equations describing the basic evolutionary effects of the motion of the system, which are the equations of its secular motion, we determine by averaging of equation (3.12) with respect to ν:

$$\frac{d\theta}{d\nu} = 0, \quad \frac{d\psi}{d\nu} = \frac{1}{2}N_0^* \cos\theta, \quad \alpha = (\psi - \psi_0)\cos\theta.$$

(3.13)

From equations (3.12), (3.13) it is possible to make the conclusion that the value of the amplitude of the longitudinal oscillations for a motion both without losing and with losing the connection does not change qualitatively the character of the evolution of the parameters of the motion of the system and determines only the velocity of the precession of the moment of momentum vector in secular motion and the amplitude of its deviations from uniform precession.

Coinciding with equations (3.11), (3.12) the equations of the first approximation describing changes of orientation of the moment of momentum vector of the relative motion of the system may be obtained by using also other methods of averaging. In fact, if we average the equations of perturbed motion in the form (2.76) only with respect to the variable φ^0 but not with respect to the variable w_3 then the equations take the form

$$\frac{d\theta}{d\varphi^0} = \tilde{N}_1 \sin\theta \cos(\nu - \psi) \sin(\nu - \psi),$$

$$\frac{d\psi}{d\varphi^0} = \tilde{N}_1 \cos\theta \sin^2(\nu - \psi),$$

$$\frac{dL}{d\varphi^0} = 0, \quad \frac{db_v}{d\varphi^0} = \frac{d(1/v)}{d\varphi^0} F_1 a_v \varphi^0,$$

$$\frac{dw_3}{d\varphi^0} = \frac{2\pi}{w_{12}} - \left[\frac{db_v}{d\varphi^0}\left(\Phi_3(w_3) - \frac{\partial a_v}{\partial b_v}\right)\right]\left(b_v \frac{d\Phi_3}{dw_3}\right)^{-1},$$

(3.14)

where

$$\tilde{N}_1 = -\frac{3}{2}\frac{\mu}{p^3}\frac{1}{\nu^4 L^2}(1 + e\cos\nu)^3.$$

We pass in equations (3.14) to differentiation with respect to t. Thus we take into account that the last two equations (the equations of longitudinal oscillations) are equivalent by virtue of their deduction to equation

$$\ddot{r} - \frac{L^2}{r^3} = -T + F_{1av\varphi^0} \tag{3.15}$$

and, therefore, with the help of the replacement of the variable it is possible to pass to the representation $r = a - b\Phi_2(w_2)$ and appropriate equations in a form (2.75).

Averaging the equations with respect to w_2 we finally obtain

$$\frac{d\psi}{dt} = N_2 \cos\theta \sin^2(\nu - \psi),$$

$$\frac{d\theta}{dt} = N_2 \sin\theta \cos(\nu - \psi) \sin(\nu - \psi),$$

$$\frac{dL}{dt} = 0, \quad \frac{db}{dt} = 0, \tag{3.16}$$

where

$$N_2 = -\frac{3}{2}\frac{\mu}{p^3}\frac{\tilde{r}^2}{L}(1 + e\cos\nu)^3, \quad \tilde{r}^2 = \frac{1}{2\pi}\int_0^{2\pi}(a - b\Phi_2(w_2))^2 \, dw_2.$$

It is not difficult to check with the help of replacement of variable of integration that $\tilde{r}^2 \equiv r^{*4}/r_*^2$

Thus, the use of the equations of perturbed motion in the form of (2.68), (2.75) and the application of the operator of averaging with respect to the variables φ and ω_2 to them results in the equations of the first approximation. However, such an approach, in comparison with averaging of the equations of the kind (2.76) with respect to the variables φ^0 and ω_3, is not acceptable for investigations of resonant modes of motion and does not give ε_2 approximations for an angle of pure rotation of the system.

Application of the operator of averaging, with respect to the angular variables φ and ω_1, to the equations of the kind (2.68), (2.74) results in the equations distinguished from (3.12) only in the length of the bar of the equivalent dumb-bell: instead of r^{*4}/r_*^2 it is $a^2 + b^2/2$. Performed calculations show that significant deviations of the solution of the averaged equations from the non-averaged equations occur here.

In Fig. 3.4 and Fig. 3.5 the characteristics of the deviations of the solutions of the averaged equations from non-averaged ones are represented. The calculations were carried out for the following parameters: orbit parameters: $e = 0.2$, $p = 7885$ km; system parameters: $d = 1000$ m, $c_m/d = 5\,\mathrm{s}^{-2}$; initial conditions of motion: $L/d^2 = 0.1\,\mathrm{s}^{-1}$; $r = 1079.27$ m, $\dot{r} = 0$, $\psi = \pi/4, \theta = \pi/6$, $\nu = \varphi = 0$.

For such parameters the change r in the unperturbed motion occurs between values $r_1 = 500$ m and $r_2 = 1079.27$ m and the period of longitudinal oscillations is equal to $\omega_{11} = 10.09$ s.

In Fig. 3.4 changes of angles ψ (Fig. 3.4a) and θ (Fig. 3.4b) on the initial interval of time are represented. From Fig. 3.4 it is visible that deviations of the solution of the averaged equations (3.6), (3.10) (line "Av") from the solution of the exact equations (line "Nonav") are almost periodic and do not exceed 0.000035 rad. The calculations on a longer interval of time do not reveal the change of the characteristics of the deviations. In Fig. 3.4 also the deviation of the solution averaged with the help of the operator of averaging with respect to the variables φ and ω_1 of the equations (2.68), (2.74) (line "Av*") from the solution of the exact equations is visible.

In Fig. 3.5 solutions of the equations of the first approximation (3.12) (line "Nonav"), the solution averaged with respect to φ and ω_1 of the equations with length of the bar of the equivalent dumb-bell $\sqrt{a^2 + b^2/2}$ (line "Nonav*") and the solution of the equations of secular motion (3.13) (line "Av") are represented. It is visible that the solutions of the equations of first approximation and the equations of secular motion have the character similar to the appropriate solutions considered above in the example of motion of the system (Fig. 3.4). It is visible on the initial interval of time that the deviation of the solution of the averaged equations for the length of the bar $\sqrt{a^2 + b^2/2}$ from the solution of the equations of first approximation for the angle θ has periodic character, reaching the value 0.00017 rad (Fig. 3.5a), and that for the

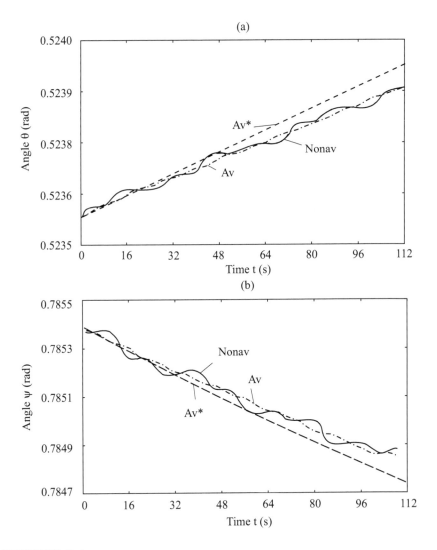

FIGURE 3.4
Variations of angles ϑ and ψ on the time initial interval.

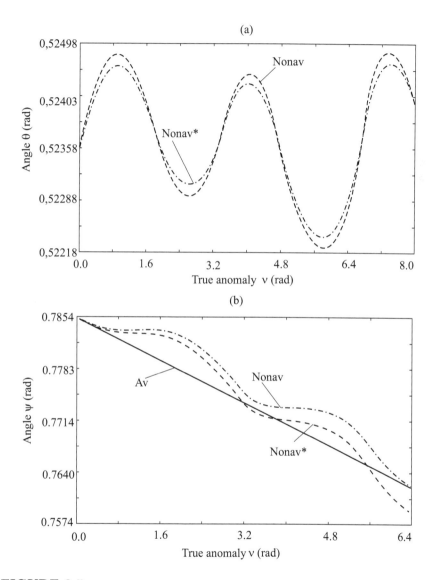

FIGURE 3.5
Deviation of angles ϑ and ψ from their values at uniform precession.

angle ψ this deviation monotonously grows approximately by 0.035 rad for one period (Fig. 3.5b).

3.1.5 Slow rotation of the system

We assume now that the angular velocity of rotation of the system about the mass centre is close to the value of angular velocity of the orbital motion, i.e., the ratio of the second to the first is not a small value:

$$\sqrt{\frac{\mu}{p^3} \frac{r^2}{L}} \sim 1.$$

This mode of motion corresponds to relative slow rotation of the system or to its oscillations about the local vertical. Concerning the velocity of change of the phase of longitudinal oscillations we assume that its value, as before, essentially exceeds the angular velocity of the orbital motion, and now, therefore, also the angular velocity of the relative motion:

$$\frac{L}{r^2} \bigg/ \frac{2\pi}{\omega_{11}} = \varepsilon_4 \ll 1.$$

For the research of this motion of the system, use of the equations of the perturbed motion represented by (2.68), (2.75) and application of the operator of averaging with respect to the angular variable ω_2 to them is most expedient. The equations of first approximation with respect to parameter ε_4, constructed in that way, look like

$$\dot{\psi} = \frac{N_3 \sin\varphi}{L \sin\theta}(\vec{e}_1, \vec{e}_R)(\vec{e}_3, \vec{e}_R),$$

$$\dot{\theta} = \frac{N_3}{L}\cos\varphi(\vec{e}_1, \vec{e}_R)(\vec{e}_3, \vec{e}_R),$$

$$\dot{L} = N_3(\vec{e}_1, \vec{e}_R)(\vec{e}_2, \vec{e}_R),$$

$$\dot{\varphi} = \frac{L}{r_{av}^2} - \dot{\psi}\cos\theta,$$

$$\dot{b} = 0, \quad \dot{\nu} = \sqrt{\frac{\mu}{p^3}}(1 + e\cos\nu)^3, \qquad (3.17)$$

where

$$N_3 = 3\frac{\mu}{p^3}\tilde{r}_{av}^2(1 + e\cos\nu)^3, \quad \tilde{r}_{av}^2 = \frac{1}{2\pi}\int_0^{2\pi}(a - b\Phi_2(\omega_2))^2\, d\omega_2,$$

$$\frac{1}{r_{av}^2} = \frac{1}{2\pi}\int_0^{2\pi}\frac{d\omega_2}{(a - b\Phi_2(\omega_2))^2}.$$

Equations (3.17) are obtained by averaging of the equations (2.68), (2.75) with respect to w_2 after the perturbing forces (3.2) were substituted.

The equations (3.17) are close to the equations of motion of the dumb-bell in the Newtonian field of forces. The difference is only that the values $\tilde{r}_{av}^2$ and r_{av}^2 are not equal. This difference is determined by the fact that the influence of the Newtonian field of forces in first approximation is equivalent to the influence on the dumb-bell with the length of the bar $\tilde{r}_{av}$ and the average angular velocity is equal to the angular velocity of motion of the dumb-bell with length of the bar r_{av}.

Example. We consider a simple example that explains what we said. Let the orbit of the mass centre be circular ($e = 0$), and the motion of the system occurs in the plane of orbit $\theta = 0$. Besides we assume that $\Phi_2(w_2) \equiv \cos(w_2)$. This assumption, in particular, corresponds to the motion of two point masses connected by a linear spring.

The equations of first approximation for the considered example we can obtain from equation (3.17):

$$\frac{dL}{dt} = -\frac{3}{2}\left(a^2 + \frac{b^2}{2}\right)\sin\lambda,$$

$$\frac{d\lambda}{dt} = \frac{L}{a^2\left(1 - \left(\frac{b}{a}\right)^2\right)^{3/2}} - \sqrt{\frac{\mu}{p^3}}, \tag{3.18}$$

$$\frac{db}{dt} = 0,$$

where $\lambda = \varphi - \nu + \psi$ is the angle between $\vec{e}_r$ and $\vec{e}_R$. From the equations (3.18) we obtain the equation of the change of the angle λ

$$\ddot{\lambda} + \frac{3}{2}\frac{\mu}{p^3}z_k \sin 2\lambda = 0, \tag{3.19}$$

which has the first integral

$$h_\lambda = \lambda'^2 - 3z_k \cos^2\lambda, \tag{3.20}$$

where

$$z_k = \left(1 + 0.5\left(\frac{b}{a}\right)^2\right)\left(1 - \left(\frac{b}{a}\right)^2\right)^{-3/2},$$

the prime designates differentiation with respect to ν. Equation (3.20) differs from the equation of the corresponding motion of the dumb-bell [1] only by the presence of the multiplier z_k at $\cos^2\lambda$. If $(b/a)^2$ is negligible small, the equations of first approximation of the system and the equations of motion of the dumb-bell coincide.

3.1.6 Energy dissipation due to the tether material

We assume that the damping properties of the tether caused by internal dissipation of energy in its material are described through the equivalent of "viscous friction" [25, 58, 117]. Then the perturbing accelerations of the dissipative forces in relative motion of the system look like

$$\vec{F}_d = \delta \chi \sqrt{\frac{c_m}{d}} \dot{r} \vec{e}_r, \tag{3.21}$$

where χ is the dimensionless coefficient of damping, δ is defined in the same way as in formula (2.23).

Concerning the value of the damping coefficient χ we assume similarly as in [25, 117] that it is a small value $\chi \ll 1$.

As experimental research [11, 86] shows, the assumptions made above of the damping properties of the tether are reasonable for many materials used in the formation of flexible elastic connections (for cables, tethers, and springs).

We consider [5, 122, 87] the influence of dissipation of energy in the material of the connection in the motion of the system where it is quickly rotating in the central Newtonian field of forces: $\varepsilon_2 \ll 1$.

3.1.7 Essentially non-linear longitudinal oscillations

For the investigation of the motion it is necessary to use the equations of perturbed motion in form (2.68), (2.75). As it was shown in subsection 3.1.2, the application of the operator of averaging with respect to variable w_2 and φ to these equations is correct and results in the construction of the equations of the first approximation.

The equations of the first approximation constructed by such a method look like

$$\dot{\theta} = N_2 \sin\theta \cos(\nu - \psi)\sin(\nu - \psi),$$

$$\dot{\psi} = N_2 \cos\theta \sin^2(\nu - \psi),$$

$$\dot{L} = 0,$$

$$\dot{b} = -\left(\frac{2\pi}{\omega_{11}}\right)^2 \frac{\chi b^2 \sqrt{c_m/d}}{\dfrac{\partial V(r_2)}{\partial r_2}\left(1 + \dfrac{\partial a}{\partial b}\right)} J, \tag{3.22}$$

$$\dot{a} = \frac{\partial a}{\partial b}\dot{b},$$

$$\dot{\nu} = \sqrt{\frac{\mu}{p^3}}(1 + e\cos\nu)^2,$$

where

$$N_2 = -\frac{3}{2} \frac{\mu}{p^3} \frac{\tilde{r}^2}{L} (1 + e\cos\nu)^3, \tilde{r}^2 = \frac{1}{2\pi} \int\limits_0^{2\pi} (a - b\Phi_2(\omega_2))^2 \, d\omega_2,$$

$$J = \frac{1}{2\pi} \int\limits_0^{2\pi} \delta \left(\frac{d\Phi_2(\omega_2)}{d\omega_2}\right)^2 \, d\omega_2.$$

From equations (3.22) it follows that in first approximation the dissipation of the energy in the material of the tether does not change the value of the moment of momentum and the character of evolution of its orientation. From (3.22) follows also that the external Newtonian field of forces practically does not render an influence on the process of attenuation of the longitudinal oscillations and the calculation of changes of parameters a and b in the first approximation is carried out independently of the change of the other variables of the system (3.22).

We note that the accuracy of equations (3.22) for the variables θ, ψ and L is of the order of ε_2 and for the variables a and b of the order of the value χ.

The integration of equations (3.22) is connected with certain peculiarities caused by fact that the integrals determining $\tilde{r}^2$ and J are not constant here and are dependent on the slowly varying variables a and b. If the calculation of the dependence of $\tilde{r}^2$ on a and b does not cause basic difficulties, since the parameters a and b can be taken out of the integral, for the calculation of J for the mode of motion with loss of tension it is obviously not possible to determine the dependence of J on a and b, since these parameters determine the borders of integration:

$$J = \frac{1}{2\pi} \int\limits_0^{2\pi} \delta \left(\frac{d\Phi_2(\omega_2)}{d\omega_2}\right)^2 \, d\omega_2 = \frac{1}{2\pi} \int\limits_{\gamma_2}^{2\pi-\gamma_2} \left(\frac{d\Phi_2(\omega_2)}{d\omega_2}\right)^2 \, d\omega_2, \qquad (3.23)$$

where γ_2 is found from the condition

$$a - b\Phi_2(\gamma_2) = d. \qquad (3.24)$$

Therefore, direct use of the equations (3.22) for the calculation of the process of attenuation of longitudinal oscillations is impossible in the motion with loss of tension.

We calculate the derivative of J with respect to time. By virtue of (3.23), (3.24) this is equal to

$$\dot{J} = \frac{1}{\pi} \left(\frac{d\Phi_2(\gamma_2)}{d\gamma_2}\right)^2 \dot{\gamma}_2 = \frac{\dot{a} - \dot{b}\Phi_2(\gamma_2)}{\pi b} \frac{d\Phi_2(\gamma_2)}{d\gamma_2} =$$

$$\frac{\dot{a} - \dot{b}\frac{a-d}{b}}{\pi b} \cdot \frac{d\Phi_2(\gamma_2)}{d\gamma_2}. \qquad (3.25)$$

Such as $\Phi_2(\cdot)$ is determined for initial values a and $b - a_0$, b_0,

$$\frac{d\Phi_2(\gamma_2)}{d\gamma_2} = \frac{\omega_{11}}{2\pi b_0}\sqrt{2V(a_0 + b_0) - 2V\left(a_0 - b_0\frac{a - d}{b}\right)}. \qquad (3.26)$$

The initial value of J is determined by integration (3.23) for $a = a_0$, $b = b_0$.

Thus, it is necessary to carry out the calculation in first approximation of the changes of parameters of the longitudinal oscillations at motion with loss of tension by integration of the following system of equations:

$$\dot{b} = -\left(\frac{2\pi}{\omega_{11}}\right)^2 \frac{b^2\chi}{\sqrt{cm/d}}\frac{\partial V(a + b)}{\partial(a + b)}\left(1 + \frac{\partial a}{\partial b}\right)J,$$

$$\dot{a} = \frac{\partial a}{\partial b}\dot{b}, \qquad (3.27)$$

$$\dot{J} = \dot{b}\frac{\dfrac{\partial a}{\partial b} + \dfrac{d - a}{b}}{2\pi^2 b^2}\frac{\omega_{11}}{b_0}\sqrt{2V(a_0 + b_0) - 2V\left(a_0 - b_0\frac{a - d}{b}\right)}.$$

Calculations show that on the time interval of the order $\left(\chi\sqrt{c_m/d}\right)^{-1}$ the solutions of equations (3.27) have high accuracy, and their absolute deviations from the solutions of the complete equations have the order of χb_0.

However, on a longer interval of time the accuracy of the solutions of the equations (3.27) is deteriorated, thus monotonous increase of the deviations of these solutions from the solutions of the exact equations takes place. This increase of the deviations is connected with the non-linear character of oscillations, namely, with change of the period of oscillations ω_{11} and the kind of function $\Phi_2(\cdot)$ for the reduction of the amplitude of longitudinal oscillations. Therefore for a more exact calculation of the changes of parameters of oscillations a and b on the equations (3.27) on a long interval of time it is required to recalculate periodically the period ω_{11} of the oscillations and the function $d\Phi_2(\gamma_2)/d\gamma_2$ for new values a, b.

We note that even in the case of recalculation of the period of oscillations ω_{11} and $d\Phi_2(\gamma_2)/d\gamma_2$ in each step of integration the integration of the equations (3.27) is much easier than the integration of the exact equations of motion. This is connected with fact that equations (3.27) do not contain discretely varying parameters and the step size of integration in view of their smoothness can be significantly larger than the step size of integration of the exact equations.

3.1.8   Linear tether stretching

The equations of first approximation become significantly simpler in the for practice important case when the stiffness of the tether is so large that the value of stretching of the tether is small in comparison with its equilibrium length $r - d/r_0 \ll 1$ at $r > d$ (d is nominal length of the connection). In this case according to the formulas of unperturbed motion [57] in motion with tense tether $\Phi_2(\omega_2)$ is very close to $\cos(kt)$, $\Phi_2(\cdot) \approx \cos(\cdot)$, $d\omega_2 = kd\,t$, and the period of longitudinal oscillations with large accuracy is determined by the finite formulas.

Hence, according to the equations (2.79) and formulas (2.39)–(2.42) equations of the modification of energy of the longitudinal oscillations, averaged with respect to φ over one period of oscillations, can be written as follows:

$$\dot{h} = -kb^* \sin \omega_2 (\chi \sqrt{c_m/d}\, b^* \sin \omega_2 + F_{ct}) \quad \forall \omega_2 \in [\gamma, 2\pi - \gamma]$$

and

$$\dot{h} = 0 \quad \forall \omega_2 \in [0, \gamma] \cup [2\pi - \gamma, 2\pi].$$

Averaging the equation of change of h with respect to ω_2 and taking into account that

$$\dot{h} = \dot{b}\frac{\partial V(r_2)}{\partial r_2}(1 + \frac{\partial a}{\partial b}),$$

we obtain that in this case the equations of the first approximation for longitudinal oscillations look like:

$$\dot{b} = -\frac{k}{\omega_{11}}b^{*2}\frac{\chi\sqrt{c_m/d}}{\dfrac{\partial V(a+b)}{\partial(a+b)}\left(1 + \dfrac{\partial a}{\partial b}\right)}(\pi - \gamma_1 + \frac{1}{2}\sin 2\gamma_1),$$

$$\dot{a} = \frac{\partial a}{\partial b}\dot{b},$$

(3.28)

where

$$k = \sqrt{\frac{c_m}{d} + 3\frac{L^2}{r_0^4}}, \quad \omega_{11} = 2\tau_0 + \frac{2\pi - 2\gamma_1}{k}, \quad \tau_0 = \sqrt{\frac{d^2 - r_1^2}{2V(r_1)}},$$

$$b^* = \frac{r_0 - d}{\cos\gamma_1}, \quad \gamma_1 = \arctan\left(\frac{L\sqrt{d^2 - r_1^2}}{k(r_0 - d)r_1 d}\right), \quad r_1 = a - b.$$

In motion without loss of tension $\gamma_1 = 0$, $\tau_0 = 0$, $b^* = b$, $a = r_0$ [5]. The equation (3.28) is obtained from the previous equation by integration of the right-hand side with respect to $d\omega_2$ from 0 to 2π and divided by 2π, i.e., by using the operator of averaging. The equations (3.28) describe modifications of longitudinal oscillations. Modifications of orientation of rotation of the system, obviously, will be described as a first approximation as well as in the equations (3.22).

The calculation of the process of attenuation of longitudinal oscillations on a long interval of time by equations (3.28) does not require recalculation of the function $\Phi_2(\omega_2)$, and the recalculation of the period of longitudinal oscillations is determined by the final formula and easily may be carried out in each step of integration.

$$\tilde{r}^2 = \frac{1}{\omega_{11}} \left\{ \left[r_1^2 + \frac{1}{3} \left(d^2 - r_1^2 \right) \right] 2\tau_0 + \frac{r_0^2}{k} \left(2\pi + 2\gamma_1 - 4\frac{r_0}{d}\gamma_1 \right) \right\}. \qquad (3.29)$$

The equation (3.29) is obtained by integration of r^2 over the period of longitudinal oscillations. r is described by formulas (2.41), (2.42) in unperturbed motion, i.e., (3.29) is an integral of r^2 from formulas (2.41), (2.42) over one period and divided by this period.

In Fig. 3.6 results of calculations of the process of attenuation of longitudinal oscillations of the system for the following parameters are shown: $P = 7885\,\text{km}$, $e = 0.2$; $c_m/d = 5\,\text{s}^{-2}$, $\chi = 0.01$, $d = 1000\,\text{m}$; $r = 800\,\text{m}$, $L/d^2 = 0.1\,\text{s}^{-1}$, $\theta = \pi/6$, $\psi = \pi/4$, $\nu = \varphi = 0$.

It is visible that the solutions of the equations (3.27) (lines "Av;" 1 is the variation of parameter a; 2, 3 are the variations of $a+b$ and $a-b$ respectively) on an initial interval of time are good coordinated with the solutions of the complete equations of motion (line "Nonav"). Their deviations on an interval of hundred seconds do not exceed 1 m which for the amplitude of oscillations makes about one percent. In Fig. 3.6 it is visible distinctly that with growing time these deviations monotonously grow and already for $t = 200\,\text{s}$ the values of deviations reach 3 m.

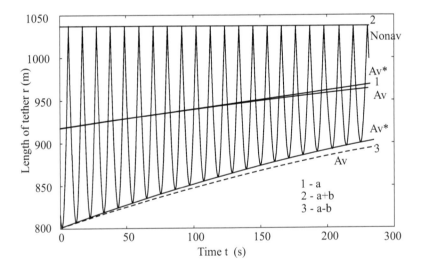

FIGURE 3.6
Damping of longitudinal oscillations for the mode of loss of tension.

Deviations of the solution of the equations (3.28) (line "Av*" in Fig. 3.6) from the solutions of the complete equations do not exceed 1% from the value of the amplitude of oscillations on all intervals of time for which calculations were performed. The increase of these deviations is absent.

3.1.9   Averaging with respect to the phases of oscillations of the unperturbed motion

Good accuracy is also given by the equations of first approximation, constructed with use of the equations of perturbed motion (2.68), (2.74), to which the operators of averaging with respect to the averaged phases of oscillations of unperturbed motion [87] are applied.

We introduce notations: $\tilde{\omega}_{11}$ is the time of motion when the system is connected for one period of oscillations; $2\tau_0$ is time of motion without connection for one period of oscillations. Therefore, the average frequency of motion when the system is connected is equal to $\omega_1^c = (2\pi - 2\gamma)/\tilde{\omega}_{11}$ and for losing connection $\omega_1^f = \gamma/\tau_0$ where $\gamma = \arccos \dfrac{a-d}{b}$.

Then it is possible to present the operator of smoothing along the generating solution with large accuracy as

$$\frac{1}{\omega_{11}} \int_0^{\omega_{11}} F\left(\bar{x}, \omega_1(C,t)\right) dt \approx \frac{\tilde{\omega}_{11}}{\omega_{11}(2\pi - 2\gamma)} \int_\gamma^{2\pi-\gamma} F\left(\bar{x}, \omega_1^c t\right) d\omega_1^c t+$$

$$\frac{\tau_0}{2\omega_{11}\gamma} \int_{-\gamma}^{\gamma} F\left(\bar{x}, \omega_1^f t\right) d\omega_1^f t, \tag{3.30}$$

where $\bar{x}$ is the vector of the slow variable and C is an arbitrary constant.

By carrying out averaging of the equations of perturbed motion with respect to the variable φ and the variable ω_1 according to the stated program we obtain the equations of first approximation

$$\dot{b} = -\chi \sqrt{\frac{c_m}{d}} \frac{\tilde{\omega}_{11}}{\omega_{11}(2\pi - 2\gamma)} \left\{ \frac{c_m}{d} b \left[4(a-d)(\pi - \gamma - \sin\gamma) + \right.\right.$$

$$b\left(\pi - \gamma - \frac{\sin 2\gamma}{2}\right) \Big] + L^2 \left[\frac{1}{(a+b)^2} + \frac{2b\sin\gamma}{d(a^2-b^2)} - \right.$$

$$\left.\left. \frac{4a}{(a^2-b^2)^{3/2}} \arctan\frac{(a-b)\cot\gamma/2}{(a^2-b^2)^{1/2}} \right] \right\} \times$$

$$\left[\frac{\partial V(a+b)}{\partial a} \left(1 + \frac{\partial a}{\partial b}\right) \right]^{-1},$$

$$\dot{\nu} = \sqrt{\frac{\mu}{p^3}} (1 + e\cos\nu)^2,$$

$$\dot{\theta} = N_2^* \cos(\nu - \psi)\sin(\nu - \psi)\sin\theta,$$

$$\dot{\psi} = N_2^* \cos\theta \sin^2(\nu - \psi),$$

$$\dot{L} = 0, \dot{a} = \frac{\partial a}{\partial b}\dot{b}, \qquad (3.31)$$

$$\tilde{\omega}_{11} = 2\int_d^{a+b} \frac{dr}{\sqrt{2V(a+b) - 2V(r)}}, \qquad N_2^* = -\frac{3}{2}\frac{\mu}{p^3}\frac{\tilde{r}^{*2}}{L}(1 + e\cos\nu)^3,$$

$$\tilde{r}^{*2} = a^2 + b^2/2 + \frac{1}{\omega_{11}}\left(\frac{\tau_0}{\gamma} - \frac{\tilde{\omega}_{11}}{2\pi - 2\gamma}\right)\left(b^2\frac{\sin 2\gamma}{2} + 4ab\sin\gamma\right).$$

In Figs 3.7, 3.8 results of computations of the motion of the system for the following parameters are presented: $P = 9000\,\mathrm{km}$, $e = 0.4$; $c_m/d = 100\,\mathrm{s}^{-2}$, $\chi = 0.02$, $d = 1000\,\mathrm{m}$; $r = 1000\,\mathrm{m}$, $\dot{r} = 9\,\mathrm{m\,s}^{-1}$, $L/d^2 = 0.1\,\mathrm{s}^{-1}$, $\theta = 0.1745\,\mathrm{rad}$, $\psi = \pi/4$, $\nu = \varphi = 0$.

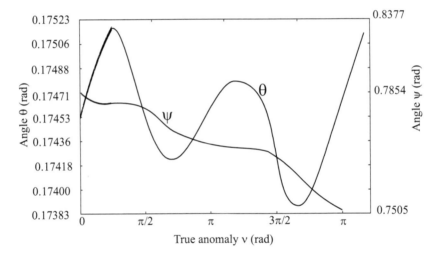

FIGURE 3.7
Variations of ϑ and ψ angles.

In Fig. 3.7 graphs of the angles θ and ψ are presented. Deviations of the solutions of equations (3.31) (broken curves in Fig. 3.7) from the solutions of the exact equations (continuous curves in Fig. 3.7) for angles θ, ψ are smaller than 0.000035 rad. In Fig. 3.9 the change of distance between bodies is shown. Deviations of the solutions of the equations (3.31) (broken curves in Fig. 3.8; 1 denotes the change of parameter a; 2, 3 denote the change $a + b$ and $a - b$ respectively) to the solutions of the exact equations (full curve) is less than 1% from the value of the amplitude of oscillations, i.e., smaller than 0.02 m.

3.1.10   Phase of slow evolution

The amplitude of longitudinal oscillations of the system according to the equations of first approximation monotonously decreases to zero. Moreover, often

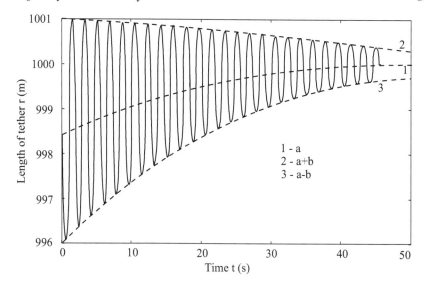

FIGURE 3.8

Damping of longitudinal oscillations.

this process goes on quite quickly. Therefore the motion of the system during attenuation of the longitudinal oscillations may be considered as the transitional mode to stationary motion, in which the amplitude of free elastic oscillations is equal to zero. From the equations of first approximation follows that in the transitive mode of motion the dissipation of energy in the material of the connection does not change qualitatively the character of the evolution of the vector of the moment of momentum.

In the stationary mode of motion (stage of slow evolution) in first approximation with respect to the value of ε_2 the motion of the system coincides with the motion of a dumb-bell with length of the bar equal to r_0, and the value of the moment of momentum is constant (equation (3.3)). Moreover, it is obvious that the simultaneous influence of dissipative forces of the tether and the Newtonian field of forces on the rotating system results in a dissipation of energy of its rotation. We estimate the reduction of the value and the possible changes of orientation of the vector of the moment of momentum caused by the dissipation of energy in the material of the tether.

We consider the influence of dissipation of energy in the material of the tether on the evolution of the parameters of motion of the system in the stationary mode of motion with an accuracy up to and including the second order of smallness [92].

The algorithm of researches for such kind of problems is traditional and consists of the following: the approximated solutions for the forced elastic oscillations as functions of the variable describing the motion of the system are constructed and then with the help of the method of averaging equations of the first approximation are constructed, based on which the analysis is carried out. Thus, the second approximation with respect to small parameters for the motion of all system is not constructed here and only in the second approximation the influence of dissipation of energy in the material of the tether on evolution of the motion of the system is investigated. The correctness of such an algorithm of research requires that the second approximation with respect to the small parameters of motion of the rigid system (the dumb-bell system) should in a negligible small way differ from the motion described by the equations of first approximation. In the other words, the correctness of the algorithm requires that for the rigid system the first approximation should describe all basic laws of motion. The reliance that this is so is given to us by the Kolmogorov–Arnold–Moser theory. By its application to the motion of a symmetric rigid body on a circular orbit [18] and certainly by numerous calculations.

3.1.10.1   First approximation for longitudinal oscillations

Since the value of r_0 is determined by equality of centrifugal and elastic forces in unperturbed motion, and in the considered mode of motion the relation of amplitude of elastic oscillations to the length of the tether is proportional to the small value ε_2, the changes of length of the tether are described with accuracy up to and including the first order of smallness by the equation

$$\ddot{z} + k^2 z + \xi \dot{z} = \frac{\mu}{R^3} r_0 \left(3(\vec{e}_1, \vec{e}_R)^2 - 1 \right) + \frac{2L_0 L_1}{r_0^3}, \tag{3.32}$$

where

$$z = r - r_0, \quad k = \sqrt{\frac{c_m}{d} + 3\frac{L^2}{r_0^4}}, \quad \xi = \chi \sqrt{\frac{c_m}{d}},$$

$\dfrac{2L_0 L_1}{r_0^3}$ is the value of change of the centrifugal accelerations for the rotation of the tether in the gravitational field considered with an accuracy to the first order of smallness from the equation of change of L:

$$\dot{L} = 3\frac{\mu}{R^3} r_0^2 (\vec{e}_1, \vec{e}_R)(\vec{e}_2, \vec{e}_R). \tag{3.33}$$

Thus, in equation (3.32) oscillations of the length of the tether which are caused both by the change of the forces acting along the line of the tether and the moment of the gravitational forces, varying the velocity of rotations of the system, are taken into account.

We allocate on the right-hand side of equations (3.32), (3.33) the fast variable φ. For this purpose we use the following representations

$$(\vec{e}_1, \vec{e}_R)(\vec{e}_2, \vec{e}_R) = 0.5(\alpha_2 \cos 2\varphi - \alpha_1 \sin 2\varphi),$$

$$3(\vec{e}_1, \vec{e}_R)^2 - 1 = 1.5(\alpha_1 \cos 2\varphi + \alpha_2 \sin 2\varphi + \alpha_3 - 2/3),$$

$$\alpha_1 = \cos^2(\nu - \psi) - \cos^2\theta \sin^2(\nu - \psi),$$

$$\alpha_2 = \cos\theta \sin 2(\nu - \psi),$$

$$\alpha_3 = \cos^2\theta \sin^2(\nu - \psi) + \cos^2(\nu - \psi).$$

The determination of the longitudinal oscillations of the tether in first approximation corresponds to the following mechanical reasoning: since in the considered motion φ is the single fast variable, for the construction of the approached solution it is possible that one does not consider the change of other variables for one period of change of φ, i.e., it is possible to consider ψ, θ, L, ν in the equations (3.32), (3.33) as constant, and $\varphi = \dfrac{L}{r_0^2}(t - t_0)$.

In fact, the solution of equation (3.33) in first approximation we find according to the method of averaging (equation of the first approximation (3.3)). Then

$$\frac{L_1^2}{r_0^3} = \frac{3}{2}\frac{\mu}{R^3}r_0(\alpha_1 \cos 2\varphi + \alpha_2 \sin 2\varphi).$$

The first approximation for the forced longitudinal oscillations of system described by equation (3.32) is constructed on the basis of the following statement [92].

Statement. Let the elastic oscillations of system be described by equations

$$\ddot{x} + k^2 x + 2\xi\dot{x} = \varepsilon_1 F(y)\sin(\omega t + t_0),$$
$$\dot{y} = \varepsilon_2 Y(y, \omega t), \tag{3.34}$$

where ε_1, ε_2 are small parameters in the sense that $\varepsilon_1/k^2 \ll 1$ $\varepsilon_1/\omega^2 = \varepsilon_2/\omega = \varepsilon \ll 1$; the value $k^2 - \omega^2$ has the order k^2 or ω^2, i.e., the system is far from the resonance 1:1; F, Y are quite smooth functions of their arguments. Then the forced elastic oscillations of the system in the first approximation with respect to the small parameter ε coincide with the oscillations described by the function $\varepsilon_1 F(y)x_0$ where x_0 is the forced oscillation of the linear system

$$\ddot{x}_0 + k^2 x_0 + 2\xi\dot{x}_0 = \sin(\omega t + t_0). \tag{3.35}$$

Proof. We pass in (3.34) to the dimensionless "time" $\tau = \omega t$:

$$x'' + \frac{k^2}{\omega^2}x + 2\frac{\xi}{\omega}x' = \varepsilon F(y)\sin(\tau + t_0), \quad y' = \varepsilon Y(y, \tau),$$

and we make the replacement of variables

$$x = \varepsilon F(y)x_0 + s,$$

$$X' = \varepsilon F(y)x_0' + s_1.$$

Then

$$s'' + \frac{k^2}{\omega^2}s + 2\frac{\xi}{\omega}s' =$$

$$-\varepsilon^2 \frac{dF}{dy}\left[2Yx_0' + x_0\left(2\frac{\xi}{\omega}Y + Y'\right)\right] - \varepsilon^3 \frac{d^2F}{dy^2}Y^2x_0.$$

Since we are interested in the forced oscillations, in view of formula (3.35) for the solution of the linear system the statement is proved.

It is also possible to assert that the true forced elastic oscillations of the system (3.34) differ from $\varepsilon_1 F(y)x_0$ by a conditionally periodic term whose amplitude has the order ε^2.

Therefore, it is easy to obtain with accuracy up to the first order of smallness that

$$z = 3\frac{\mu}{R^3}\frac{r_0}{k_2^2}(A\cos 2\alpha + B\sin 2\alpha + D), \tag{3.36}$$

$$A = \alpha_1 - \xi_1\alpha_2, \quad B = \alpha_2 + \xi_1\alpha_1, \quad D = \frac{1}{2}\frac{k_2^2}{k^2}\left(\alpha_3 - \frac{2}{3}\right),$$

where

$$k_2^2 = k_1^2(1 + \xi_1^2), \quad k_1^2 = \frac{c_m}{d} - \frac{L^2}{r_0^4}, \quad \xi_1 = 2\frac{\xi L}{r_0^2 k_1^2},$$

and it is supposed that k_1^2, k_2^2 have the order of smallness k^2 or L^2/r_0^4, i.e., the system is not in a resonance 1:2.

3.1.10.2   Laws of motion

Substituting in (2.68), (3.2) the found expression $r_0 + z$ and performing averaging of the equation with respect to the angular variable φ we obtain the equations of first approximation describing the influence of elastic-dissipative properties of the tether with accuracy up to and including the order ε_2

$$\dot{\psi} = N_1\cos\theta\sin^2(\nu - \psi) - N_d\left[\cos\theta\sin^2(\nu - \psi)(\alpha_3 + 2D) + 0.5\xi_1\sin 2(\nu - \psi)\alpha_3\right],$$

$$\dot{\theta} = 0.5N_1\sin\theta\sin 2(\nu - \psi) - N_d\sin\theta\left[\sin 2(\nu - \psi)\times (0.5\alpha_3 + D) - \xi_1\alpha_3\cos\theta\sin^2(\nu - \psi)\right],$$

$$\dot{L} = -N_dL\xi_1(\alpha_2^2 + \alpha_1^2),$$

$$\dot{\nu} = \sqrt{\frac{\mu}{p^3}}(1 + e\cos\nu)^2, \tag{3.37}$$

where

$$N_d = \frac{9}{2}\left(\frac{\mu}{R^3}\right)^2\frac{r_0^2}{L}\frac{1}{k_2^2}.$$

From the equations (3.37) it follows that the specific velocity of decrease of L is negligible small. We are interested in the qualitative aspect of this finding.

Let us consider the influences of dissipative forces on the motion of the system. For this purpose we write out the equations of change of the moment of momentum retaining only the members of equations (3.37), reflecting the influence of energy dissipation in the material of the tether on its evolution:

$$\dot{\psi} = -0.5 N_d \xi_1 \sin 2(\nu - \psi) \alpha_3,$$

$$\dot{\theta} = 0.5 N_d \xi_1 \sin 2\theta \sin^2(\nu - \psi) \alpha_3,$$

$$\dot{L} = -N_d L \xi_1 (\alpha_2^2 + \alpha_1^2). \tag{3.38}$$

Taking into account that $\alpha_2^2 + \alpha_1^2 = \alpha_3^2$,

$$\frac{\partial \alpha_3^2}{\partial \theta} = -\sin 2\theta \sin^2(\nu - \psi) \alpha_3,$$

$$\frac{\partial \alpha_3^2}{\partial \psi} = 2 \sin^2 \theta \sin 2(\nu - \psi) \alpha_3,$$

it is possible to make the conclusion that the influence of the dissipative forces in each instant of time tries to arrange the moment of momentum vector in a position minimizing the velocity of its decrease. And the direction of the influence of the dissipative forces is close to the direction of the "steepest descent" for the value α_3^2. Since

$$\gamma^2 = (\vec{e}_3, \vec{e}_R)^2 = \sin^2 \theta \sin^2(\nu - \psi),$$

$$\frac{\partial \gamma^2}{\partial \theta} = \sin 2\theta \sin^2(\nu - \psi),$$

$$\frac{\partial \gamma^2}{\partial \psi} = -\sin^2 \theta \sin 2(\nu - \psi),$$

by virtue of (3.38) the dissipative forces in each moment of time try to combine the moment of momentum with the vector $\vec{R}$, i.e., try to transfer the rotation of the system in a plane that is perpendicular to the vector $\vec{R}$. This orientation of rotation of the system corresponds to the minimum of dissipation of energy of its rotation.

We consider the specific power of the dissipative forces

$$N = -\xi \dot{z}^2. \tag{3.39}$$

Using the expression for z (3.36) we find with an accuracy up to ε_2^2 the specific power of dissipative forces averaged for the tether period of rotation around of the mass centre

$$\langle N \rangle_\varphi = -9 \left(\frac{\mu}{R^3} \right)^2 L \xi_1 \frac{\alpha_3^2}{k_2^2}. \tag{3.40}$$

Since the average specific power of the dissipative forces and the velocity of decrease of the moment of momentum are proportional to α_3^2, by virtue of the previous analysis it is possible to make the conclusion that the influence of the dissipative forces in each instant of time tries to reduce the absolute work of these forces and in the end tries to transfer the system into a state corresponding for the given mode of motion to a possible minimum of the absolute value of their work.

We obtain the equations of the basic evolutionary effects by performing in equations (3.37) differentiation with respect to ν and with their subsequent averaging with respect to ν

$$\frac{d\psi}{d\nu} = \frac{1}{2}N_0 \cos\theta - I_d \left\{ \cos\theta \left[\left(1 + \frac{k_2^2}{k^2}\right) \beta_1 - 4\frac{k_2^2}{k^2} \left(\frac{2}{3} + 2e^2 - \right.\right.\right.$$
$$\left.\left. e^2 \cos 2\psi\right)\right] - 3\xi_1 e^2 \sin 2\psi (1 + \cos^2\theta) \bigg\},$$

$$\frac{d\theta}{d\nu} = I_d \sin\theta \left[3e^2 \sin 2\psi \beta_2 + \xi_1 \cos\theta \beta_1 \right],$$

$$\frac{dL}{d\nu} = -I_d L \xi_1 \left[(1 + 3e^2)(3\cos^4\theta + 3 + 2\cos^2\theta) + \right.$$
$$\left. 6e^2 \cos 2\psi (1 - \cos^4\theta) \right], \qquad (3.41)$$

where

$$\beta_1 = (1 + 3e^2)(3\cos^2\theta + 1) - 6e^2\cos^2\theta\cos 2\psi,$$

$$\beta_2 = \left(1 + \frac{k_2^2}{k^2}\right)(1 + \cos^2\theta) - \frac{4}{3}\frac{k_2^2}{k^2},$$

$$I_d = \frac{9}{16}\left(\frac{\mu}{p^3}\right)^{\frac{3}{2}} \frac{r_0^2}{L}\frac{1}{k_2^2},$$

N_0 is the same expression as in equations (3.4). The terms proportional e^4 are here omitted as they do not add any qualitative differences to the solution of the equations.

From equations (3.41) it is visible that under the action of dissipation of energy in the material of (3.40) the tether the moment of momentum of the system tries to lie in the plane of orbit ($\theta \to \pi/2$). It is also an interesting fact that for θ close to $\pi/2$ the action of the dissipative forces traces an elliptic shape of the orbit and tries to arrange the plane of rotation of the tether perpendicularly to the radius-vector of the pericentre of the orbit ($\psi \to \pm\pi/2$). This effect also corresponds to the tendency of the system to lower the velocity of decrease of the moment of momentum of the relative motion.

Let us estimate the averaged specific power of the dissipative forces for one period of the orbital motion. For this purpose we use the following scheme:

$$\langle\langle N\rangle_\varphi\rangle_\nu = \frac{1}{T_\nu}\int_0^{T_\nu}\langle N\rangle_\varphi dt \approx \frac{1}{2\pi}\int_0^{2\pi}\langle N\rangle_\varphi\frac{R^2}{\sqrt{\mu p}}d\nu, \quad (3.42)$$

where T_ν is the period of the orbital motion.

Then

$$\langle\langle N\rangle_\varphi\rangle_\nu = -\frac{9}{8}\left(\frac{\mu}{p^3}\right)^3\xi_1\frac{L}{k_2^2}\left[(3\cos\theta + 3 + 2\cos^2\theta)\left(1+\right.\right.$$

$$3e^2 + \frac{3}{8}e^4\right) + \cos 2\psi(1 - cos^4\theta)(6e^2 + e^4) +$$

$$\left.\frac{1}{16}e^4\cos 4\psi(\cos^4\theta + sin^2\theta)\right]. \quad (3.43)$$

The obtained formula (the terms proportional e^4 are kept here) as well as (3.40) is similar to the formula of velocity of decrease of the moment of momentum, i.e., the attempt of the system to reduce the output of energy naturally coincides with the effort to reduce the absolute value of work of the dissipative forces in the considered case. The further analysis of the laws of motion of the system under the influence of dissipative forces is carried out in the research of the influence of aerodynamic forces on the relative motion of the system and in the research of translational-rotary motion.

The character of the changes of θ and L under the influence of dissipation of energy in the material of the tether is shown in Fig. 3.9. The calculations were carried out for the following values of parameters: $P = 6621\,\text{km}$, $e = 0.2$; $c_m/d = 0.25\,\text{s}^{-2}$, $\chi = 0.05$, $d = 10000\,\text{m}$; $L/d^2 = 0.05\,\text{s}^{-1}$, $\theta = 0.1745\,\text{rad}$, $\psi = \pi/6$, $\nu = \varphi = 0$.

3.1.11 Influence of aerodynamic forces

It is natural to introduce aerodynamic influences on the system as forces acting on each of the bodies of the system [4, 25, 43, 117]

$$\vec{F}_{ai} = -\dot{\vec{R}}_i\left|\dot{\vec{R}}_i\right|k_{ai}m_i, \quad (3.44)$$

where $k_{ai} = \rho_i c_{xi}S_i/2m_i$, ρ_i is the density of the atmosphere, S_i is the projected area, c_{xi} is the aerodynamic factor of resistance $(i = 1, 2)$.

We make the usual assumptions for this case [3, 5, 6, 16, 43]: the disturbing influence of the aerodynamic forces is small, i.e., $R\cdot\max\{k_{a1}, k_{a2}\} = \varepsilon_5 \ll 1$, the value of the relative velocity of the motion of the system is much smaller than the velocity of orbital motion $|\dot{\vec{r}}| \ll |\dot{\vec{R}}|$.

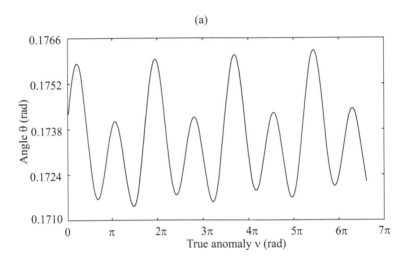

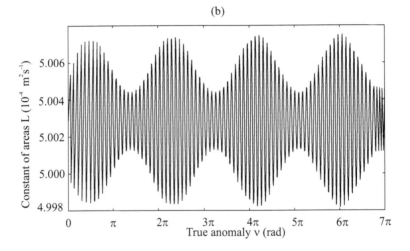

FIGURE 3.9
Character of changing ϑ and L in a stationary mode of motion.

Then the influence on the relative motion of the system up to the terms of order $k_{ai}|\vec{r}|^2$ looks like

$$\vec{F}_a = (k_{a1} - k_{a2})|\dot{\vec{R}}|\dot{\vec{R}} - I\left[|\dot{\vec{R}}|\dot{\vec{r}} + (\dot{\vec{R}}, \dot{\vec{r}})\dot{\vec{R}}/|\dot{\vec{R}}|\right]. \tag{3.45}$$

Here

$$I = (k_{a2}m_1 + k_{a1}m_2)/M, \quad |\dot{\vec{R}}| = \sqrt{\frac{\mu}{p}}\sqrt{1 + 2e\cos\nu + e^2}.$$

The second term in (3.45) determines the dissipative component of the aerodynamic accelerations.

Let us consider the influence of aerodynamic accelerations determined by the (3.45) on the relative motion of a quickly rotating system $\dfrac{\mu}{p^3}\dfrac{r^4}{L^2} = \varepsilon_2 \ll 1$. We assume that the amplitude of longitudinal oscillations of the system is small in comparison with the length of the tether:

$$\left(\frac{b}{r_0}\right)^2 = \varepsilon_3 \ll 1.$$

3.1.12 Equations of first approximation

The equations of first approximation under the influence of perturbing accelerations of the Newtonian field of forces (3.1), dissipation of energy in the material of the tether (3.21) and aerodynamic forces (3.45) we construct by the way of averaging the equations in the form (2.68), (2.79) along the generating solution (2.31), (2.34). Assuming that the change of density of the environment ρ_i does not depend on the relative motion of the system ($\rho_1 = \rho_2$) we obtain

$$\dot{\psi} = \frac{I\mu}{2|\dot{\vec{R}}|p}\beta_{a1}\beta_{a2} + N_1\cos\theta\sin^2(\nu - \psi),$$

$$\dot{\theta} = \frac{I\mu}{2|\dot{\vec{R}}|p}\cos\theta\sin\theta\beta_{a1}^2 + N_1\sin\theta\cos(\nu - \psi)\sin(\nu - \psi),$$

$$\dot{L} = -L\left[|\dot{\vec{R}}|I + \frac{I\mu}{2|\dot{\vec{R}}|p}\left(\beta_{a2}^2 + \cos^2\theta\beta_{a1}^2\right)\right],$$

$$\dot{b} = -\frac{b}{2}\left[\chi\sqrt{\frac{c_m}{d}} + |\dot{\vec{R}}|I + \frac{I\mu}{2|\dot{\vec{R}}|p}\left(\beta_{a2}^2 + \cos^2\theta\beta_{a1}^2\right) + \right.$$

$$\left. \frac{1}{2}\frac{L\dot{L}}{r_0^4k^2}\left(12\frac{L^2}{r_0^4k^2} - 3\right)\right],$$

$$\dot{r}_0 = \frac{2L\dot{L}}{r_0^3k^2},$$

$$\dot{\nu} = \sqrt{\frac{\mu}{p^3}}(1 + e\cos\nu)^2, \tag{3.46}$$

where $\beta_{a1} = \cos(\nu - \psi) + e\cos\psi$, $\quad \beta_{a2} = \sin(\nu - \psi) - e\sin\psi$, and N_1 is the same expression as in equation (3.3).

As it was to be expected, the change of the moment of momentum in the considered mode of motion does not depend in the first approximation on the longitudinal oscillations of the system. The longitudinal oscillations of the system damp out due to the dissipation of energy both in the material of the tether and from the aerodynamic resistance.

Thus, the analysis of the evolution of the orientation of the system and the influence of the perturbing effects on it may be carried out on the model of the dumb-bell with slowly varying length of the bar equal to r_0 and determined by the equality of the centrifugal and elastic forces

$$\frac{c_m}{d}(r_0 - d) = \frac{L^2}{r_0^3}. \tag{3.47}$$

The perturbing aerodynamic accelerations in the first approximation influence the relative motion of the system only by the dissipative component. Their conservative component results only in small almost periodic oscillations of the system.

3.1.13   Influence of dissipative aerodynamic forces

We carry out the analysis of the influence of the dissipative component of the aerodynamic forces on the motion of the system. For this purpose we write out the equations of the change of the moment of momentum retaining only those terms of the equations (3.46) reflecting the influence of the aerodynamic resistance to its evolution:

$$\dot{\psi} = \frac{I\mu}{2|\dot{\vec{R}}|p}\beta_{a1}\beta_{a2},$$

$$\dot{\theta} = \frac{I\mu}{2|\dot{\vec{R}}|p}\cos\theta\sin\theta\beta_{a1}^2,$$

$$\dot{L} = -L\left[|\dot{\vec{R}}|I + \frac{I\mu}{2|\dot{\vec{R}}|p}\left(\beta_{a2}^2 + \cos^2\theta\beta_{a1}^2\right)\right]. \tag{3.48}$$

Since

$$\frac{\partial\tilde{\beta}_a}{\partial\theta} = -2\cos\theta\sin\theta\beta_{a1}^2,$$

$$\frac{\partial\tilde{\beta}_a}{\partial\psi} = -2\sin^2\theta\beta_{a1}\beta_{a2},$$

where

$$\tilde{\beta}_a = \beta_{a2}^2 + \cos^2\theta\beta_{a1}^2,$$

it is possible to make the conclusion that the influence of the dissipative component of the aerodynamic forces in each instant of time tries to arrange the moment of momentum in the state which minimizes the velocity of its decrease. And the direction of these influences is close to the direction "of the steepest descent" for the value $\tilde{\beta}_a$.

Comparison of the obtained results with conclusions about the influence of energy dissipation in the material of connection on the motion of the system allows to introduce the hypothesis [92] that the influences of dissipative forces of various physical nature are directed to transferring the motion of the system into a state corresponding to the minimum loss of energy.

This supposition is similar to the known principle [14] about the tendency of material systems to avoid friction. However, in [14] this principle is considered as the resultant tendency in the motion of systems: the velocities of motions with stipulated dissipation of the energy, under action of dissipative forces become equal to zero. For the considered system "to avoid friction" is possible only by reaching the position of relative equilibrium corresponding to the arrangement of the system along the local vertical. The effect of dissipative forces in the considered case in each instant is directed to the change of parameters of motion, pursuant to this principle, i.e., even if it sounds paradoxically, in the direction of decreasing the dissipation of energy.

From equations (3.48) it follows that the influence of aerodynamic forces tries to arrange the vector of the moment of momentum in tp plane of orbit $(\theta \to \pi/2)$. This effect of the influence of the aerodynamic forces occurs for any form of the orbit of the mass centre. Taking into account that

$$(\vec{e}_3, \dot{\vec{R}}) = -\sqrt{\mu}p\sin\theta\beta_{a1}, \quad \frac{\partial \sin^2\theta\beta_{a1}^2}{\partial\theta} = \sin\theta\cos\theta\beta_{a1}^2, \quad \frac{\partial\beta_{a1}^2}{\partial\psi} = 2\beta_{a1}\beta_{a2},$$

it is possible by virtue of (3.48) also to make the conclusion that in each moment of time under the influence of the aerodynamic resistance the vector of the moment of momentum tries to approach the direction that is collinear to the vector of the velocity of the mass centre. This direction corresponds to the minimum of the velocity for the reduction of the value of the moment of momentum.

Velocities of changes of angles of orientation of the moment of momentum and the velocity for the reduction of its specific value are proportional to the value $\sqrt{\dfrac{\mu}{p^3}}\varepsilon_5$.

3.1.14 Basic laws of evolution of motion

For the determination of the main evolutionary effects of the motion of the system under the influence of perturbations we carry out averaging of the equations (3.46) with respect to variable ν. For this purpose, as before, we first pass in the equations (3.46) to differentiation with respect to ν. The equations describing the main evolutionary effects of the change of the three-dimensional orientation of the system look like:

$$\frac{d\psi}{d\nu} = -\frac{1}{2}(I_1 - I_2)\sin\psi\cos\psi + \frac{1}{2}N_0\cos\theta,$$

$$\frac{d\theta}{d\nu} = \frac{1}{2}\cos\theta\sin\theta\left(I_1\cos^2\psi + I_2\sin^2\psi\right),$$

$$\frac{dL}{d\nu} = -L\left[I_3 + \frac{1}{2}I_1\left(\sin^2\psi + \cos^2\theta\cos^2\psi\right)\right.$$
$$\left. + \frac{1}{2}I_2\left(\cos^2\psi + \cos^2\theta\sin^2\psi\right)\right], \tag{3.49}$$

where

$$I_1 = \frac{p}{2\pi}\int_0^{2\pi}\frac{I(1 + e\cos\nu)^{-2}(\cos\nu + e)^2}{\sqrt{1 + 2e\cos\nu + e^2}}\,d\nu,$$

$$I_2 = \frac{p}{2\pi}\int_0^{2\pi}\frac{I(1 + e\cos\nu)^{-2}\sin^2\nu}{\sqrt{1 + 2e\cos\nu + e^2}}\,d\nu,$$

$$I_3 = \frac{p}{2\pi}\int_0^{2\pi}\frac{I\sqrt{1 + 2e\cos\nu + e^2}}{(1 + e\cos\nu)^2}\,d\nu,$$

N_0 is the same as in (3.4).

For the circular orbit $I_1 = I_2 = 0.5I_3$ and the equations for θ and L are easily integrated:

$$\tan\theta = \tan\theta_0\exp\left\{\frac{1}{2}I_1(\nu - \nu_0)\right\},$$

$$\tag{3.50}$$

$$L = L_0\exp\left\{-3\frac{1}{2}I_1(\nu - \nu_0)\right\}\frac{\cos\theta_0}{\cos\theta}.$$

From expressions (3.50) as well as from equations (3.48) follows that on the circular orbit the vector of moment of momentum under the action of the dissipative component of the aerodynamic forces tries to lie in the plane of the orbit. This conclusion is distinct from the conclusion [16, 23] formulated for a symmetric rigid body, according to which the influence of the dissipative component of the aerodynamic forces does not change direction of the vector of moment of momentum on a circular orbit. The difference of conclusions is caused by the discrepancy used in [16, 23] of the formulas approximating the dissipative moment of aerodynamic forces and the moment of these forces acting on the dumb-bell. Simultaneously the above formulated conclusion corresponds to results of work [83] about the influence of the dissipative component of aerodynamic forces on the motion of a spherical satellite around the mass centre. The numerical solution of the exact equations of motion confirms the correctness of the above made conclusion and the results of the work [22] confirm the generality of the obtained results for the motion of rigid bodies on the orbit.

In Fig. 3.10 the motion of the vector of moment of momentum on the unit sphere is presented under the joint influence of perturbations of the Newtonian field of forces and aerodynamic forces on a circular orbit of motion of the mass centre. The trajectory seems to wind up on unit sphere from the pole to the equator with increasing density.

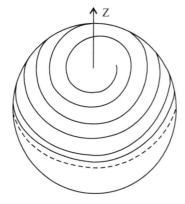

FIGURE 3.10

Trajectory of the moment of momentum vector on the unit sphere.

For the motion of the system on an elliptic orbit there is an additional effect caused by the action of aerodynamic forces. Depending on the sign of the difference $I_1 - I_2$ the vector of the moment of momentum under the action of aerodynamic forces tries to coincide either with the direction of the tangent to the orbit in its pericentre ($I_1 > I_2$), or with the direction collinear to the radius-vector of the orbit in its pericentre ($I_2 > I_1$). As it is visible from calculations for models in which the density of the environment depends only on the distance between the mass centre and the attractive centre $I_1 > I_2$. However, for models taking into account other factors of change of density of the environment it can also appear that $I_2 > I_1$.

Independently of the sign of the value $I_1 - I_2$ on the basis of equations (3.49) it is possible to formulate the general law of the influence of aerodynamic forces on the relative motion of the system: under the action of aerodynamic forces the system tries to get into the position in which the value of the vector of moment of momentum of the relative motion decreases with minimum velocity, i.e., tries to get the position of the minimum aerodynamic resistance to the

relative motion. This action is caused by the dissipative component of the
aerodynamic forces.

We consider laws of evolution of the motion of the system on an elliptic
orbit under joint influence of a Newtonian field of forces and aerodynamic
forces. We assume that $I_1 - I_2 > 0$.

We consider first the evolution of the orientation of the system under
the assumption of constancy of L. This case is interesting for the projects of
cable systems in which the preservation of constant velocity of rotation or the
increase of the velocity of rotation of the system is supposed. It is easy to see
that this can be achieved by introduction of additional accelerations into the
system which are directed along the axis Oy. Thus the equations of evolution
of the angles ψ and θ do not change.

For constant L the value $|N_0 \cos \theta|$ monotonously decreases in time. There-
fore, for any initial conditions of motion a moment of time exists when the
aerodynamic influences on the precession of the vector of moment of momentum exceeds the influences of the Newtonian field of forces ($I_1 - I_2 >
|N_0 \cos \theta|$). The further part of the trajectory has "aerodynamic" character,
i.e., the vector of the moment of momentum tries to reach the direction of the
tangent to the orbit in its pericentre.

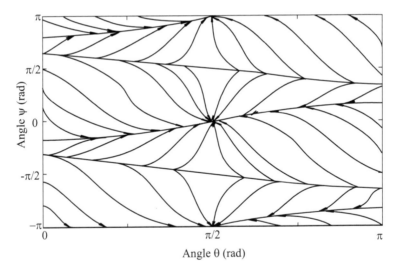

FIGURE 3.11
Phase portrait of angles ϑ and ψ variations.

The phase portrait of change of angles ψ, θ for constant L for $I_1 - I_2 > |N_0|$
is shown in Fig. 3.11. The equation of lines 1, 2, 3, 4 looks like:

$$(I_1 - I_2) \sin \psi \cos \psi = N_0 \cos \theta. \tag{3.51}$$

In Fig. 3.12 for the considered case the character of possible trajectories of the vector on unit sphere is presented. The dashed lines 1, 2, 3, 4 correspond to similar lines in the phase portrait, the continuous lines show possible trajectories of the moment of momentum vector.

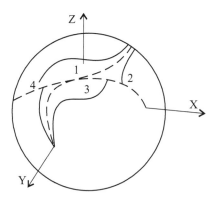

FIGURE 3.12
Possible trajectories of the moment of momentum vector on the unit sphere.

For the case of the increase of the value of the moment of momentum vector the character of its evolution is similar.

If the evolution of the value of the moment of momentum vector occurs according to equations (3.49), the value $|N_0 \cos \theta|$ grows strictly monotonously with time, and therefore, the influence of the central field of forces grows with time. Thus, if in an initial moment of time the condition $|N_0 \cos \theta| > I_1 - I_2$ is fulfilled, the character of evolution of the moment of momentum vector differs little from its evolution on a circular orbit.

For the fulfilment of the condition $|N_0 \cos \theta| < I_1 - I_2$ in an initial moment of time, i.e., when the aerodynamic influences on the precession of the moment of momentum vector exceed the influences of the central field of forces depending on the initial value ψ two variants of motion are possible. Either ψ has an initial value such that

$$(I_1 - I_2) \sin \psi \cos \psi < N_0 \cos \theta, \qquad (3.52)$$

the trajectory of evolution of the vector of the moment of momentum has a point of change of the direction of the precession determined by equation (3.50) where L already is not constant and changes according to equations (3.49). In this case the initial segment of the trajectory has "aerodynamic" character (Fig. 3.13). On this segment the trajectory comes nearer to one of the lines determined by equation (3.51) (corresponding to lines 1 or 3 for the case of constant L) and having touched it changes the sign of precession. Later the slow transition into rotatory motion around the normal to the plane of the orbit occurs.

Or if the initial value of ψ is such that the condition (3.52) is fulfilled, the change of the direction of precession does not occur. However, in this case the initial segment of the trajectory differs significantly from the trajectory on a circular orbit (Fig. 3.14) too. On the initial segment of the trajectory the vector of the moment of momentum comes closer to one of the corresponding lines determined by equation (3.51) from opposite sides but does not touch this line. Then, as in the case of fulfilment of condition (3.52) the slow transition into rotatory motion around the normal to the plane of the orbit occurs.

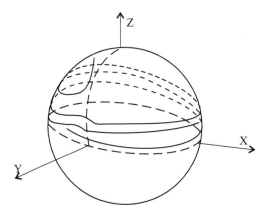

FIGURE 3.13
Trajectories of the moment of momentum vector on the unit sphere. Case of changing of precession sign.

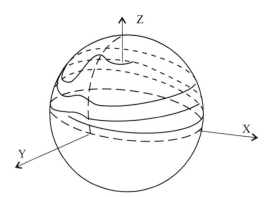

FIGURE 3.14
Trajectories of the moment of momentum vector on the unit sphere. Precession does not change the sign.

3.1.15 Influence of other perturbing factors

Relative motion of the system under the assumption that the trajectory of the mass centre is an unperturbed Keplerian orbit was considered above. However, such influences on the system as the influences caused by the deviation of the gravitational field of the Earth from the central Newtonian field and influences of environmental forces of resistance result in a change of the trajectory of motion of the mass centre. At the same time it is often of practical interest that it is possible to consider, with large accuracy, that the motion of the mass centre of the system is independent of the system's relative motion. Thus under appropriate assumptions concerning the smallness of the values $(r/R)^2$ and $|\dot{\vec{r}}|/|\dot{\vec{R}}|$ it is possible to consider that the influence of the gravitational field of forces and the environmental resistance on the orbital motion of the system does not depend on its relative motion. In these cases the research of relative motion is carried out also within the framework of the limited statement of the problem, the difference of which consists only, that the known trajectory of motion of the mass centre is a perturbed Keplerian orbit, i.e., a Keplerian orbit with slowly varying parameters. It is obvious that the technique of research of the relative motion of the system is the same in this case.

It is most expedient to take into account the evolution of the angular orbital parameters i, Ω, ω_π with the help of the equations of perturbed motion along the evolving orbit.

For such an approach expressions of influences on the system depending on the radius-vector of the mass centre and its velocity (expressions of influences of the central field of forces and aerodynamic forces) do not change. The influence of evolutions of angular orbital parameters is described by additional terms in the right-hand parts of the equations.

The evolution of the parameters of the orbit p and e is taken into account parametrically, i.e., in the corresponding formulas p and e are already considered not to be constant but as slowly varying parameters.

Thus, the calculation of the evolution of the orbit of the system investigating its relative motion within the framework of the limited statement of the problem does not result in basic difficulties and is carried out precisely in the same way as the research of the relative motion of a rigid body [16, 18]. Thus, the scheme of conclusions from the equations of the first approximation, the expressions on the right-hand sides of these equations, corresponding to the influences of the considered perturbations do not change, and consequently, do not vary the conclusions about the influence of the considered perturbations on the relative motion of the system. In particular, the motion of a system under the influence of the Newtonian field of forces and the influence on the evolution of the orbit in the first approximation coincides with the similar motion of the dumb-bell the qualitative analysis of which can be found in [18].

We point out that the influence of the evolution of the orbit on the change of orientation of the rotational motion of the cable system of two bodies about the orbit for many Earth orbits is one of the basic perturbing factors. And hence, although the earlier determined influences of other perturbing factors do not vary, their effect on the motion of the system can be suppressed in the evolution of the orbit of the mass centre.

We consider, for example, the rotational motion of the system with an elastic-dissipative tether in view of the evolution of the orbit under the action of a non-central field of gravitation of the Earth. From (3.41), (2.48) and (2.50) it is easy to obtain that the equations of secular motion about the evolving orbit have the following form:

$$\frac{d\tau_1}{d\nu} = \frac{1}{2}N_0\cos\theta - I_d\cos\theta\left[\left(1+\frac{k_2^2}{k^2}\right)\left(3\cos^2\theta+1\right)-\frac{8}{3}\frac{k_2^2}{k^2}\right] -$$

$$\varepsilon_{or}\cos i\left[\cot\theta\sin i\cos\tau_1+\cos i\right],$$

$$\frac{d\theta}{d\nu} = \frac{1}{2}I_d\xi_1\sin 2\theta\left(\cos^2\theta+1\right)+\varepsilon_{or}\cos i\sin i\sin\tau_1,$$

$$\frac{dL}{d\nu} = -I_d L\xi_1\left(3\cos^4\theta+3+2\cos^2\theta\right). \tag{3.53}$$

Here the terms proportional to e^2, e^4 are omitted and it is assumed that in secular motion only the longitude of the ascending unit of the orbit $\dot{\Omega} = \varepsilon_{or}\cos i$, and the argument of the pericentre ω_π, $\tau_1 = \omega_\pi + \psi$, vary; ε_{or} is a small parameter.

As ε_{or} usually significantly exceeds I_d, from equations (3.53) follows that the dissipative effects in the motion of the system are possible only for orbits close to equatorial or polar ones. For other orbits the plane of orbit evolves faster than the moment of momentum of relative motion approaches it.

A similar conclusion can be made for the effect of the dissipative component of the aerodynamic forces also.

Researches of perturbed motions of the system carried out in this chapter allow to make some generalisations about the influence of perturbing forces of various physical nature on the relative motion of the system.

Let us assume that the perturbing influences have the force function $U(r,\varphi,\psi,\theta,t)$. We assume also that the initial conditions of motion and the parameters of the system are such that the angle of pure rotation of the system φ and the phase of longitudinal oscillations ω are fast variables in relation to other variables of the system, and U is periodic with respect to φ. Then it follows from the equations of perturbed motion that at any value of the amplitude of the longitudinal oscillations and independently whether the motion is with loss of tension or without it the motion of the system in absence of resonance between φ and ω coincides in the first approximation with the similar motion of the dumb-bell, the length of the bar of which is a function of a and b. This conclusion is the consequence from the following statement: in the absence of resonance, in first approximation, the value of the moment of momentum L, the amplitude of longitudinal oscillations b, and consequently the averaged length a of system are values kept constant. This statement is similar to the Laplace–Lagrange theorem in celestial mechanics.

For the influence of the dissipative forces on the relative motion of a system such as in the case with dissipation of energy in the material of the tether and for the influence of the dissipative component of the aerodynamic forces it is necessary to expect that this influence tries to arrange the system in a state which is minimizing the velocity of decrease of its moment of momentum, i.e., in a state corresponding to the minimum of dissipation of energy of the relative motion. Against the hypothesis [14] about the attempt of the material systems to avoid friction we assume that this tendency of action of dissipative forces has constant and global character, i.e., the influence of the dissipative forces in each moment of time is directed to the change of all parameters of motion such that to reduce the dissipation of the energy of motion.

In the case of smallness of the gyroscopic forces, i.e., for the motion of the system under the influence of conservative and dissipative forces, it is necessary to expect that the system tries to enter in a state corresponding to the minimum decrease of energy of the relative motion.

3.2 Interaction of translational and rotational motions

3.2.1 Equations of motion

In this section the motion of the system in a Newtonian field of forces is considered [89]. No other external forces are present $\vec{F_i} \equiv 0$. The investigations are carried out up to and including the accuracy ε_1^2, $(\varepsilon_1 = r/R)$. The perturbing accelerations of the Newtonian field of forces on the relative (equation (2.1)) and the orbital (equation (2.2)) motion of the system within this accuracy have, respectively, the form

$$\vec{F} = \frac{\mu}{R^2} \frac{r}{R} \left\{ -\vec{e}_1 + 3(\vec{e}_1, \vec{e}_R)\vec{e}_R + 3\frac{m_1 m_2}{M} \frac{r}{R} \left[(\vec{e}_1, \vec{e}_R)\vec{e}_1 + \right. \right.$$

$$\left. \left. \frac{1}{2} \left(1 - 5(\vec{e}_1, \vec{e}_R)^2 \right) \vec{e}_R \right] \right\}, \tag{3.54}$$

$$\vec{F}^* = \frac{\mu}{R^2} \left(\frac{r}{R} \right)^2 \frac{m_1 m_2}{M} \left\{ 3(\vec{e}_1, \vec{e}_R)\vec{e}_R + \frac{3}{2} \left(1 - 5(\vec{e}_1, \vec{e}_R)^2 \right) \vec{e}_R \right\}.$$

We write down the equations of perturbed motion of the system in the following form. The first group of equations of perturbed motion about an evolving orbit of the mass centre is

$$\dot{\psi} = \frac{rF_3 \sin \varphi}{L \sin \theta} + \cot \theta \left(\frac{di}{dt} \sin \tau_1 - \dot{\Omega} \sin i \cos \tau_1 \right) - \dot{\Omega} \cos i - \dot{\omega}_\pi,$$

$$\dot{\theta} = \frac{rF_3 \cos \varphi}{L} - \dot{\Omega} \sin i \sin \tau_1 - \frac{di}{dt} \cos \tau_1,$$

$$\dot{L} = rF_2, \tag{3.55}$$

$$\dot{\varphi} = \frac{L}{r^2} - \dot{\tau}_1 \cos \theta + \dot{\Omega} \left(\sin \theta \sin i \cos \tau_1 - \cos \theta \cos i \right) - \frac{di}{dt} \sin \theta \sin \tau_1,$$

where $\tau_1 = \omega_\pi + \psi$. Equation (3.55) is the equation (2.48), (2.50) taking into account (2.80).

The equation of the change of r is

$$\ddot{r} - \frac{L}{r^3} = -T + F_1, \tag{3.56}$$

and the equations of perturbed Keplerian motion are (2.69).

Equations (2.2), (3.56) in view of the various forms of the perturbed longitudinal oscillations (see section 2.4) and (2.87) form the complete system of the equations of perturbed motion for the considered problem and the methods of averaging described in Chapter 3 are perfectly applicable to them.

3.2.2   First integrals

The constant moment of momentum of the system $\vec{G}$ is equal to the sum of the moment of momentum of orbital and relative motions

$$M \sqrt{\mu p} \vec{e}_3^* + \frac{m_1 m_2}{M} L \vec{e}_3 = \vec{G}, \tag{3.57}$$

where $\vec{e}_3^*$ is the unit vector of the axis CZ.

We connect with $\vec{G}$ the non-rotating system of coordinates $C\xi\eta\zeta$ such that the axis $C\zeta$ is directed as the vector $\vec{G}$ (Fig. 3.15). Then the projections of the vectorial equality on the axes of the non-rotating coordinates system yields three first area integrals [89]:

$$\omega_\pi = \pi - \psi,$$

$$\sqrt{\mu p} \cos i + \frac{m_1 m_2}{M^2} L \cos(\theta - i) = G', \tag{3.58}$$

$$\frac{m_1 m_2}{M^2} \frac{L}{\sin i} = \frac{\sqrt{\mu p}}{\sin(\theta - i)} = \frac{G'}{\sin \theta},$$

where $G' = |\vec{G}|/M$.

The second and third equality (3.58) are relations of the triangle, formed by the vectors of moment of momentum of motions (Fig. 3.15). The first equality (3.58) follows from the construction of the systems of coordinates and means that during the motion the axis of rotation of the tethered system about the mass centre is located in the plane normal to the line of nodes of the orbit of the mass centre that coincides with the third general law of Cassini [18].

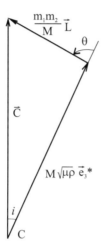

FIGURE 3.15
Triangle formed by moment of momentum vectors.

From equality (3.57), (3.58) it follows that the limited statement of the problem, i.e., the assumption about the independence of the orbit of the mass centre from the relative motion requires not only the fulfilment of the condition for r/R but also smallness of the relation of the value of moment of momentum of the relative motion to the value of moment of momentum of the orbital motion. And this relation should be a value of higher order of smallness than the considered perturbing influences.

In the considered case of gravitational influences the limited statement of the problem requires to fulfill the following condition:

$$\frac{m_1 m_2}{M^2} \frac{L}{\sqrt{\mu p}} \Big/ \frac{r}{R} \approx \frac{m_1 m_2}{M^2} \frac{r}{R} \frac{\omega}{\omega_0} \ll 1$$

where ω, ω_0 are values of the angular velocities of orbital and relative motion, respectively.

From equality (3.57), (3.58) it also follows that for the choice of the orientation of the axes of the non-rotating coordinate system, such that during the motion the inclination of the orbit is small enough, namely, that the value $\sin i$ has the order of smallness ε_1, even the small perturbations can result in essential changes of the elements of the orbit Ω and ω_π. This is connected to the degeneration of the kinematic equations for Eulerian angles for small angles of nutation.

3.2.3 Basic laws of evolution of the system

From the equations of the perturbed motion of the system and from the analysis of the influence of longitudinal oscillations on the relative motion of the system (section 1.1.3) it follows that there is no resonance in the motion of the system and if the frequency of the longitudinal oscillations significantly exceeds the average frequency of the orbital motion, the evolution of the parameters of motion of the system in first approximation coincides with their evolution for the motion of a dumb-bell with a certain (obtained) length of the bar. And in particular if the amplitude of the longitudinal oscillations is small in comparison with the length of the tether, the motion of the system in first approximation of the small parameters coincides with the motion of a dumb-bell with a length of the bar that is equal to the equilibrium length of the tether r_0.

We consider the basic evolutionary effects of the translational-rotational motion of the system, which is quickly rotating about the mass centre $\varepsilon_2 \ll 1$ $(\varepsilon_2 = \dfrac{\mu}{p^3}\dfrac{r^4}{L^2})$. For this purpose by virtue of what has been said above, it is enough to investigate the motion of the dumb-bell with length of the bar equal to r_0.

The scheme of derivation of the equations of the main evolutional effects of motion is the same, as studying the relative motion. At first, the equations of perturbed motion (3.55), (2.69) are averaged with respect to φ, taking into account expressions for a perturbing force (3.54), i.e., the averaging operator with respect to an angular variable is applied. Then, transformation to a new independent variable in the equations, which is an angle of pure rotation of orbital motion (ν) and the subsequent averaging with respect to this variable, is performed. The such constructed equations of the basic evolutionary effects of motion of the dumb-bell taking into account (3.58) look like

$$\frac{d\theta}{du} = \frac{di}{du} = \frac{dL}{du} = \frac{dp}{du} = \frac{de}{du} = 0,$$

$$\frac{d\Omega}{du} = N_0 \frac{\sin\theta\cos\theta}{\sin(\theta - i)}, \tag{3.59}$$

$$-\frac{d\psi}{du} = \frac{d\omega_\pi}{du} = -\frac{d\Omega}{du}\cos i - \frac{3}{8}\frac{m_1 m_2}{M^2}\frac{r_0^2}{p^2}(1 - \cos^2\theta).$$

Hence the basic evolutionary effect of the quickly rotating system about its mass centre consists in the rotation of the plane formed by the vectors of moment of momentum of the relative and orbital motions around the total moment of momentum vector of the system [89]. Thus one component of the angular velocity of this rotation depends neither on the masses of the bodies nor on the linear size of the system and is proportional to the ratio of angular velocities of orbital and relative motions.

3.2.4 Dissipation of energy due to the visco-elastic tether material

In section 3.1.6 influence of internal dissipation of energy in the tether material on the evolution of the parameters of relative motion within the framework of the limited formulation of the problem was investigated. And although this influence in this stage of slow evolution for most of real space systems is negligibly small, the determination of the laws of influence of dissipative forces represents general interest.

The question of evolution of extended viscous-elastic systems in the Newtonian field of forces is interesting for celestial mechanics and for the definition of general laws of motion of such systems and is the subject of permanent investigations and discussions (see, for example, [106]). The used elementary viscous-elastic system of two material points has allowed to carry out a deeper analysis of the relative motion and allows to consider general laws of translational and rotational motion. Different from the works [71, 72, 105, 115] the researches are carried out within the framework of classical mechanics of point masses that, in particular, allows to carry out the simple numerical check of the obtained results, and also the general formulation of the problem, i.e., the spatial motion of system for any orbit of the mass centre, is considered [92].

Forced longitudinal oscillations in the first approximation with respect to the small parameters ε_1, ε_2 are determined by formula (3.36). Then taking into account (3.58) the equations of perturbed motion of the system are the equations of perturbed Keplerian motion (2.29) and the equation of change of the angle φ

$$\dot{\varphi} = \frac{L}{r^2} - \dot{\Omega}\cos(\theta - i).$$

The procedure of construction of the equations of first approximation is similar to the procedure of construction of equations (3.41). The gravitational influences are taken into account up to and including the second order of smallness with respect to $(r/R)^2$. Then the equations describing the basic evolutionary effects of the translational and rotational motion of the system look like

$$\frac{di}{du} = I_{d1}\sin\theta\left\{3e^2\cos\theta\sin 2\omega_\pi\beta_2 + \right.$$

$$\left. \xi_1\left[(1+3e^2)(3+\cos^2\theta)+6e^2\cos 2\omega_\pi\right]\right\},$$

$$\frac{d\Omega}{du} = \frac{1}{2}N_0\frac{\sin\theta\cos\theta}{\sin(\theta-i)} - I_d\frac{\sin\theta}{\sin(\theta-i)}\left[\cos\theta\left(\frac{k_2^2}{k^2}+1\right)\beta_1 - \right.$$

$$\frac{4}{3}\cos\theta\frac{k_2^2}{k^2}(2+6e^2-3e^2\cos\theta\cos2\omega_\pi) +$$

$$\left. \xi_1 e^2(1+\cos^2\theta)\sin2\omega_\pi\right],$$

$$\frac{dp}{du} = -2I_{d1}p\left\{3e^2\sin^2\theta\sin2\omega_\pi\beta_2 - 4\xi_1\cos\theta\left[(1+3e^2)\times\right.\right.$$

$$\left.\left.(1+\cos^2\theta) + 1.5e^2\sin\theta\cos2\omega_\pi\right]\right\},$$

$$\frac{de}{du} = eI_{d1}\left[\sin^2\theta\sin2\omega_\pi\beta_2 + 2\xi_1\left(11+11\cos^2\theta +\right.\right.$$

$$\left.\left.7\sin^2\theta\cos2\omega_\pi\right)\right],$$

$$\frac{d\omega_\pi}{du} = -\frac{d\Omega}{du}\cos i - \frac{3}{8}\frac{m_1m_2}{M^2}\frac{r_0^2}{p^2}(1-3\cos^2\theta) - I_{d1}\left[\ldots +\right.$$

$$\left. 6\xi_1\cos\theta\sin^2\theta\sin2\omega_\pi(3+e^2)\right]. \tag{3.60}$$

Here

$$I_{d1} = \frac{9}{16}\frac{m_1m_2}{M^2}\left(\frac{r_0}{p}\right)^2\frac{\mu}{p^3}\frac{1}{k_2^2},$$

β_1, β_2 I_d is the same as in the equations (3.41).

Here the terms proportional e^4 are omitted as they do not bring in any qualitative differences to the motion of the system. In the equation of change of ω_π a bulky expression is omitted also because in this case it does not carry interesting information.

Equations (3.60), in spite of the fact that they with relations (3.58) completely describe the evolution of the slow variables of the system, are inconvenient for the analysis as their right-hand parts depend on parameters of the relative motion L and θ, the representation of which through parameters of the orbital motion is unnecessarily bulky. The question of the connection of equations (3.60) in relation to equations (3.41) also remains open.

We construct the equations for L and θ by differentiating the relation of the triangle formed by the moment of momentum with respect to u by virtue of (3.60):

$$\sqrt{\mu p}\cos i + \frac{m_1m_2}{M^2}L\cos(\theta-i) = G_1,$$

$$\sqrt{\mu p}\sin i + \frac{m_1m_2}{M^2}L\sin(\theta-i) = 0. \tag{3.61}$$

We obtain the equations

$$\frac{d(\theta - i)}{du} = I_d \sin \theta \left[-3e^2 \sin 2\omega_\pi \beta_2 + \xi_1 \cos \theta \beta_1 \right],$$

$$\frac{dL}{du} = -I_d L \xi_1 \left[(1 + 3e^2)(3 \cos^4 \theta + 3 + 2 \cos^2 \theta) + \right.$$

$$\left. 6e^2 \cos 2\omega_\pi (1 - \cos^4 \theta) \right]. \tag{3.62}$$

Thus, the equation of change of L coincides with the analogous equation in (3.41) and the equation of change of $\theta - i$ coincides with the equation of change of θ in (3.41).

From equations (3.60), (3.62) follows that the effect of dissipative forces (the members of the equations containing the multiplier ξ_1) are directed to the following changes in the motion of the system: the moment of momentum of relative motion of the system decreases, being redistributed in the moment of momentum of orbital motion; the eccentricity of the orbit ($e \neq 0$) grows; the inclination of the orbit decreases to zero. Here it is taken into account that

$$I_{d1} \xi_1 \sin \theta = \frac{9}{8} \frac{\mu}{p^3} \frac{G_1}{p^2} \frac{\xi}{k_1^2 k_2^2} \sin i; \tag{3.63}$$

the angle of nutation θ has the tendency to reach some value $\pi/2\alpha_c$, $0 < \alpha_c < \pi/2$, and only in the end, as $\dfrac{di}{du}$ reaches zero, θ goes to $\pi/2$. I.e., the influence of dissipative forces tries to transfer the inverse rotation of the system in direct one, in which the attitude rotation has the same orientation as the orbital one.

We consider laws of evolution of the system determined by the equations of first approximation.

For $e = 0$ equations (3.60), (3.62) become significantly simpler, and the evolution of the parameters is completely determined by the influence of dissipative forces. α_c is determined from equation

$$I_{d1} \xi_1 \sin \theta = \frac{9}{8} \frac{\mu}{p^3} \frac{G_1}{p^2} \frac{\xi}{k_1^2 k_2^2} \sin i; \tag{3.64}$$

Here (3.63) and that

$$I_d \xi_1 \sin \theta = \frac{9}{8} \frac{\mu}{p^3} \frac{G_1}{p^2} \frac{\xi}{k_1^2 k_2^2} \sin(\theta - i)$$

are taken into account.

In the general case ($e \neq 0$), depending on the relation of the small values ε_1 and ε_2, various characterizations of the motion of the system are possible, since the values N_0, I_d, I_{d1} which characterize the velocities of evolution of the parameters of motion of the system have the order $\sqrt{\varepsilon_2}$, $\varepsilon_2^{3/2}$, $\varepsilon_1^2 \varepsilon_2$ respectively.

Let us consider the limiting cases. We assume $(r/p)^2 \ll I_d$, i.e., $\varepsilon_1^2 \ll \varepsilon_2^{3/2}$ or which is the same $L/\sqrt{\mu p} \ll 1$ (case of distant orbits). Then $i \ll 1$, θ tries to reach $\pi/2$, and for θ close to $\pi/2$ the pericentre of the orbit tries to appear in the plane formed by the moments of momentum of the system, i.e., motion of the system corresponds to case of motion considered above about the unperturbed trajectory of the mass centre.

In the other limiting case when $\varepsilon_1 \gg \varepsilon_2^{1/2}$, i.e., $L/\sqrt{\mu p} \gg 1$ the velocity and the direction of change of the argument of the pericentre of the orbit are different and are defined by the second term in the equation for ω_π. Here it is interesting to note that as numerical integration of the equations (3.60), (3.62) shows in the case when the moment of momentum of the relative motion exceeds in value the value of the moment of momentum of orbital motion "ejection" of the system on a hyperbolic trajectory ($e \to 1$) is possible. Moreover, for inverse rotation ($i > \pi/2$) such opportunity of "ejection" is realised for a much wider area of initial parameters than for pure rotation of the system.

In the general case, different from the motion on the unperturbed orbit, the effect of stopping the precessional motion and tracking an elliptically shaped orbit vanishes (the pericentre of the orbit leaves much faster than the moment of momentum approaches to it). Depending on the values of moment of momentum and the ratio r_0/p the character of motion of the system can have significant differences from the motion of the system about an unperturbed orbit. The research of possible motions of the system exceeds the framework of the given researches.

In most cases the main effect of evolutionary motion of the system, such as for the motion of a quickly rotating dumb-bell (3.59), consists in rotation of the plane formed by the moment of momentum of orbital and relative motions around the total moment of momentum. Moreover, when θ is not close to $\pi/2$ ($|\cos\theta| > \varepsilon_1^2$, $|\cos\theta| > \varepsilon_2^{3/2}$) ω_π is the fast variable in equations (3.60), (3.62). Therefore the research of evolution of the parameters of motion can here be carried out by averaging of equations (3.60), (3.62) with respect to ω_π. It is easy to see that the equations obtained as result of this averaging operation differ from the initial ones only by the fact that all members of the equations depending on ω_π become equal to zero.

The analysis of equations (3.60), (3.62) averaged with respect to ω_π, just as the investigation by numerical integration of equations (3.60), (3.62) for θ close to $\pi/2$ shows that the evolution of the motion of the system is determined by the influence of dissipative forces, i.e., the change of the parameters of motion of the system occurs according to the direction of influence of dissipative forces and the earlier determined effects of their influence are realised in the motion of the system.

We consider the opportunity of interpretation of the motion of the system under the action of dissipative forces [90]. It is easy to see that the expression of average power of dissipative forces (3.43) does not change. But then the influences of dissipative forces directed on the increase of the eccentricity and the transfer of the system to direct rotation are directed on increase of capacity of dissipative forces, i.e., the change of parameters of orbital motion under influence of dissipative forces cannot be explained within the framework of proposed hypothesis about aspiration of systems to avoid friction as of the current tendency. The attempt of dissipative forces to increase the eccentricity of the orbit and to transfer the rotation of the system into a direct one is opposite to the tendency to reduce the energy of the relative motion that took place in the case of the constant orbit.

At the same time these influences of dissipative forces correspond to their tendency to reduce losses of energy or to increase energy for orbital motion. Really, the increase of eccentricity corresponds to this tendency. Analysing the changes in the motion of the system in comparison with motion of the system about an unperturbed orbit it is possible to make the conclusion that the introduced differences in the change of the angle θ also are in general agreement with this tendency: its acceleration decreases for $\theta > \pi/2$, its deceleration increases for $\theta < \pi/2$. The tendency to direct rotation can be interpreted as tendency to decrease (to increase) velocity of decrease (increase) of the moment of momentum of orbital motion.

Thus, the general picture of the influence of dissipative forces on the motion of the system is formed by their tendency to reduce the output of the energy (to increase its reception) for each motion, the orbital and the relative.

In the considered case, since the evolution of the motion of the system is defined by the influence of dissipative forces the general picture of evolution of the motion develops by the tendency of each of the motions "to avoid friction," i.e., by the tendency of each of the motions — orbital and relative — to decrease the output of energy (to increase its reception) for the motion.

Certainly, the considered motion of the system is only a specific example. Here two weakly connected motions (orbital and relative) take place, the dissipation of energy is described by internal tether force and occurs, resulting from the relative motion. But the distinctly shown tendency in the direction of the influence of dissipative forces on the decrease (increase) of energy of each of the motions is interesting because the arising contradictions between the forms of motions give the chance to develop not only trivial forms motions.

3.3 Regular and chaotic motions of TSS with inextensible tether

Now we consider relative motion of two point masses (m, m_0) in orbit, connected by a massless flexible inextensible string with impact reaching connection. Such a problem was put and studied in 1969 [17, 26] as a simple model of dynamics of an orbital tether system (for example, dynamics of the space probe and a satellite connected by a cable). This model can be treated as certain dynamic billiard [19].

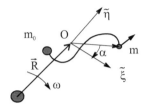

FIGURE 3.16
An orbital tether.

We consider a plain problem. Let the mass centre O of a system (m, m_0) move along a Keplerian circular orbit (Fig. 3.16). We connect orbital coordinate basis $O\xi\eta$ with the mass centre where the axis $O\xi$ is directed on tangent to circular orbit in a direction of motion, an axis $O\eta$ — in a direction of radius-vector of the mass centre O. Let l be length of a string, ω_0 be an angular velocity of orbital motion of the mass centre. We enter dimensionless time $\tau = \omega_0 t$ and dimensionless coordinates $\varrho(\xi, \eta)$ and velocities $v((\dot\xi), (\dot\eta))$ of the point m by formulas

$$\vec{\varrho} = \vec{r}\frac{m + m_0}{lm_0}; \vec{v} = \frac{\vec{V}}{\omega_0 l}, \tag{3.65}$$

where ϱ and V are dimensional coordinates and velocities. Equations of motion then look like

$$\ddot\xi + 2\dot\eta = 0, \ddot\eta - 2\dot\xi - 3\eta = 0, \varrho^2 = \xi^2 + \eta^2 \leq 1. \tag{3.66}$$

Here points mean derivatives with respect to dimensionless time, $(-2\dot\eta, 2\dot\xi)$ are components of Coriolis forces , $(0, 3\eta)$ are components of the sum of forces of a gyroscopic and gravitational gradient. It is necessary to add to (3.66) a condition of absolute-elastic (on statement) reaching connection $\varrho = 1$. This condition connects value of velocity v_+ and v_- before and after impact:

$$\vec{v}_+ = \vec{v}_- - 2(\vec{e}_r\vec{v}_-)\vec{e}_r. \tag{3.67}$$

Equations (3.66) in view of a condition of absolutely elastic reaching connection have an integral of energy

$$\dot\xi^2 + \dot\eta^2 - 3\eta^2 = h. \tag{3.68}$$

The considered problem is piece-wise integrable that facilitates research. As an example in Fig. 3.17 the family of single-link periodic trajectories (calculated on analytical formulas [27, 28]) is shown. Along a family values of energy h and period T ($T = 2\pi$ corresponds to period of orbital motion) are indicated.

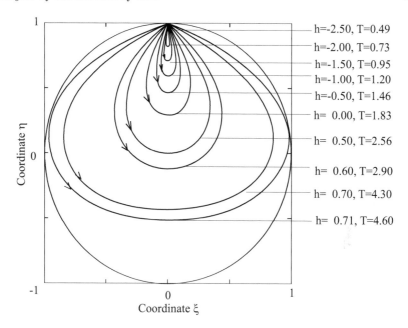

FIGURE 3.17
A set of single-link trajectories.

Mirrored family exists also reflected concerning an axis of abscissas. It is shown in [24] that all these trajectories are stable in all area $h \leq 0.71$ of the existence.

Phase space of a problem (3.66) is four-dimensional. However, at the moment of impacts one coordinate ($\varrho = 1$) is known and at the fixed value of a constant of energy h the phase space is reduced to a two-dimensional one. It allows to use a method of point map for construction of phase portraits of a problem.

Let us enter an angle α measured from a direction of motion of the mass centre of the system counter-clockwise (i.e., against a direction of motion of the mass centre (Fig. 3.16)). We choose a phase plane $(\alpha, \dot{\alpha})$ as a two-dimensional phase space at the moment of impacts. Phase portraits in this plane and their changes of aircraft attitude at change of energy h are shown in following figures (where some periodic trajectories are represented also in the configuration space (ξ, η)).

Motion values of a constant of energy has oscillatory character and has a high degree of a regularity in an interval $-3 \leq h \leq -2$. The major periodic motion (single-link) is represented in Fig. 3.17. In a phase portrait it corresponds to a stable point ("centre"). With h increase the regularity is more and more lost. Chaotic layers arise and are increased with increase h in area $-2 \leq h \leq 0$; archipelagoes of islands of multi-link trajectories arise and disappear in the chaotic sea, but stable single-link periodic motion exists always.

An example is in Figs 3.18, 3.19 ($h = -0.75$). Oscillatory motions are still in the field of $0 \leq h \leq 1$, but it is possible to see a transition to rotations also. An example is in Figs 3.20, 3.21 ($h = 0.6$). At $h > 1$ chaotic and regular rotations (Figs 3.22, 3.23 $h = 2.5$) dominate only. It is interesting that direct rotations ($\dot{\alpha} < 0$) in main are regular, but inverse ones ($\dot{\alpha} > 0$) are chaotic in main. The analysis shows that this effect is result of an action of Coriolis forces.

A phase portrait consists of regular trajectories actually all at large values h and its structure is determined by two (stable and unstable) two-link periodic motions (Figs 3.24, 3.25, $h = 25$). At $h \to \infty$ all picture aims at a set horizontal straight lines that (as well as in a billiard case in a uniform field) answers to classical kinematic circular Birkgoff billiard.

In [17] the countable family of impact-free periodic motions consisting of arcs of linked and free motions (losing connection and reaching connection happen in a impact-free way) was revealed. Initial data $(\alpha, \dot{\alpha})$ and period T of such motion's satisfy a system of transcendental equations [27, 28, 55]:

$$\tan \frac{T}{2} = \cot \alpha_0 \frac{\dot{\alpha}_0}{2\dot{\alpha}_0 - 3}; \frac{T}{2} = \frac{1}{3} \cot \alpha_0 \frac{1 - 2\dot{\alpha}_0}{2 - \dot{\alpha}_0};$$

$$\dot{\alpha}_0^2 - 2\dot{\alpha} + 3 \sin^2 \alpha_0 = 0 \tag{3.69}$$

with an energy constant

$$h = \dot{\alpha}_0^2 - 3 \sin^2 \alpha_0. \tag{3.70}$$

Two solutions of a system (3.69) give (approximately)

$$\alpha_0^{(1)} = 0.566, \dot{\alpha}_0 = 1.3706, T = 2.9028, h = 1.058, \alpha_0^{(2)} = \pi + \alpha_0^{(1)},$$

the remaining parameters are the same, as in the first solution.

Initial data of the remaining solutions are well calculated under approximate formulas

$$T = 2\pi n - \alpha_0; \alpha_0 \approx \frac{1}{6\pi n}; h \approx -\frac{1}{12\pi^2 n^2}; n = 1, 2, 3, \ldots \tag{3.71}$$

(For the second subfamily we have $\alpha_0 = \pi + 1/6\pi n$. The formulas (3.69) and (3.71) give the same result to within five digits after a point.

$$h_1 = -0.00844; h_2 = -0.00211; h_3 = -0.00094; h_4 = -0.00053.$$

The structure of circumscribed trajectories is shown in [17, 27, 28].

In [17] motion of a bodies' tether was considered at absolutely inelastic impact and it was shown that impact-free periodic trajectories (3.69) are accessible in process of impact evolution of motion. However, the probability of these events was estimated as zero. In [55] it is shown on the basis of results of [54] that this probability is distinct from zero, though it is very small ($\sim 10^{-7}$). In the same research [17] it is shown that if initial energy (3.70) of

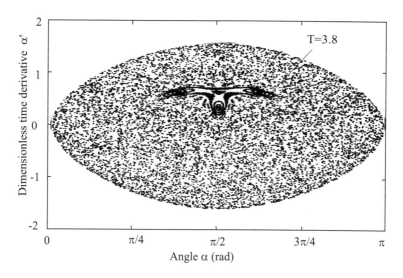

FIGURE 3.18
The phase portrait; $h = -0.75$.

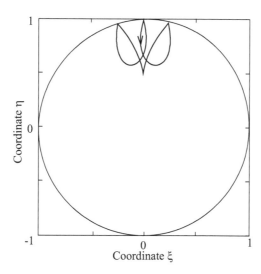

FIGURE 3.19
The three-link trajectory; $h = -0.75$.

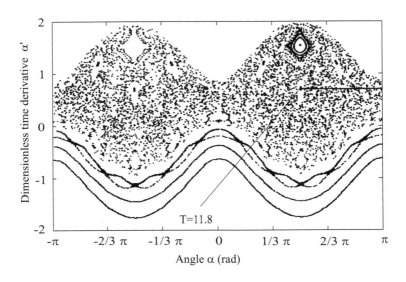

FIGURE 3.20
The phase portrait; $h = 0.6$.

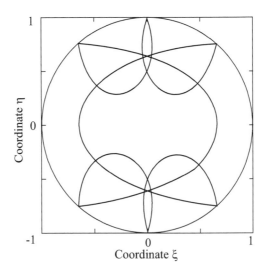

FIGURE 3.21
The six-link trajectory; $h = 0.6$.

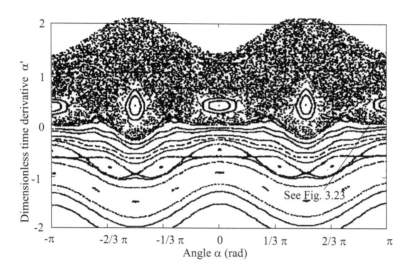

FIGURE 3.22
The phase portrait; $h = 2.5$.

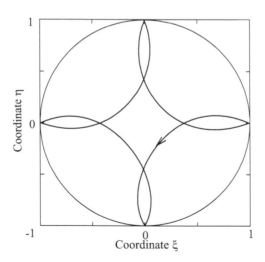

FIGURE 3.23
The fourth-link trajectory; $h = 2.5$.

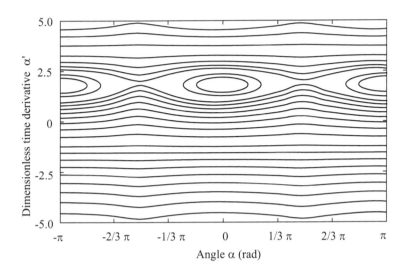

FIGURE 3.24
The phase portrait; $h = 25$.

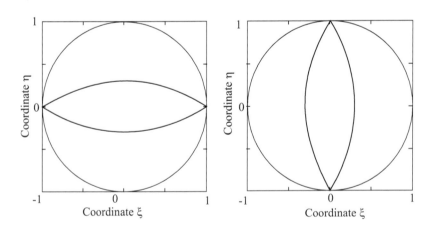

FIGURE 3.25
The two-link trajectories; stable on the left and unstable on the right; $h = 25$.

connected motion lies in an interval $-0.5 < h < 4$ a system during inevitable losing connection and impact (absolute-inelastic impact) reaching connection may evolve so that for infinite quantity of impacts it goes

on a limit cycle of periodic oscillations with amplitude 1.1346 rad and value $h = -0.5$. But it is possible also that the system will leave on one of periodic oscillations with any value h from an interval $-1.85 < h < -0.5$ for a finite number of impacts. The appropriate oscillation frequencies lie in an interval from 0.8377 rad up to 0.7715 rad. In [29, 30] the effect of aerodynamic forces of pressure on dynamics of an orbital tether of bodies is considered. Instead of equations (3.66) and an integral (3.68) equations and an integral will take place

$$\ddot{\xi} + 2\dot{\eta} + a = 0, \ddot{\eta} - 2\dot{\xi} - 3\eta = 0, \xi^2 + \eta^2 \leq 1. \tag{3.72}$$

$$\dot{\xi}^2 + \dot{\eta}^2 + 2a\xi = h. \tag{3.73}$$

The new constant parameter "a" describes an effect of aerodynamic pressure . In Figs 3.26, 3.27 the example of a phase portrait of this problem with absolutely elastic impact is shown at reaching connection (above). Centre of a regularity answers to a stable, folded, horizontally located single-link periodic trajectory. Centres of the archipelago of five islands answer to the stable five-link periodic trajectory shown below. Values of parameters: $a = 2.0, h = -1.0$.

It is visible that the aerodynamics gives absolutely new quality of trajectories: periodic trajectories as though "blown off" by an aerodynamic pressure. We note that if in a purely gravitational case (without the aerodynamics, $a = 0$) there is only one family (two subfamilies) of single-link folded periodic trajectories with "spout" in a point ($\xi = 0, \eta = 1$) or in a point ($\xi = 0, \eta = -1$) [27, 28], in view of the aerodynamics ($a \neq 0$) two such families appear. The first of these families is an oblique loop with "spout" which coordinates $x_{i0}, \eta_0, \xi_0^2 + \eta_0^2 = 1$ are located on the "left" semicircle, so the parameter ξ_0 may have any value on an interval $-1 \leq \xi_0 \leq 0$; the second family is "horizontal" loops with "spout" in a point ($\xi = -1, \eta = 0$). In [30] analytical formulas for these periodic solutions are derived and the area of their existence is investigated.

At presence of the aerodynamics families of impact-free motions exist too (however, the found families are not prolongation on parameter "a" of families (3.69), (3.71)). At small $\alpha_0, \dot{\alpha}$ these parameters and the period T of motion on a family of periodic impact-free motions are connected by ratio

$$\dot{\alpha}_0 \approx -\frac{a}{2}; \frac{2\alpha_0}{a} \approx n\pi + \arctan\frac{2\alpha_0}{a}; T \approx 4\frac{\alpha_0}{\alpha}, n = 1, 2, 3, \ldots \tag{3.74}$$

In Fig. 3.28 some of such impact-free trajectories are shown.

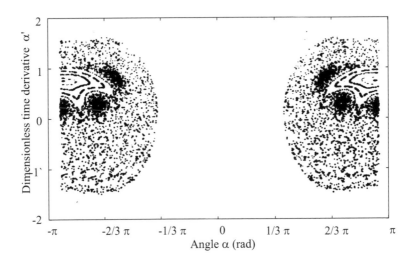

FIGURE 3.26
The phase portrait; effect of the aerodynamics.

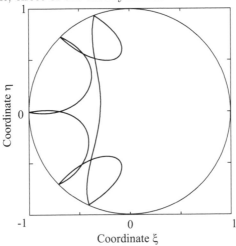

FIGURE 3.27
The five-link trajectory; effect of the aerodynamics.

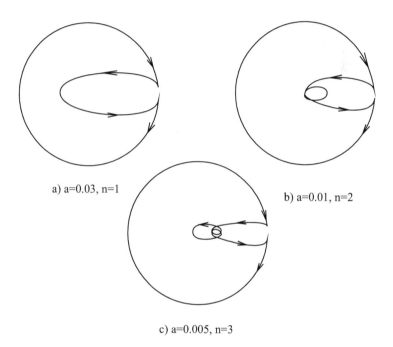

a) a=0.03, n=1

b) a=0.01, n=2

c) a=0.005, n=3

FIGURE 3.28
Effect of the aerodynamics: trajectories without impacts.

3.4   Chaotic motion of TSS with extensible tether

3.4.1   Statement of the problem

Let us consider the motion of two material points, connected by a massless linear spring. Such a model can be considered as a simple part of a computational model for studying the dynamics of a cable [25, 79], and allows to take into account oscillations of the masses of the internal degrees of freedom. We consider the motion of the system in the plane of the circular orbit of its mass centre. Then it is possible to write the equations of motion (2.68), (2.75) as follows

$$\dot{L} = -1.5\omega_0^2 r^2 \sin 2\psi, \quad \dot{\psi} = \frac{L}{r^2} - \omega_0,$$

$$\dot{b} = b\frac{\omega_0^2}{k} F_b(L, r), \quad \dot{\omega} = k + \frac{\omega_0^2}{k} F_\omega(L, r), \tag{3.75}$$

where $r = a + b\cos\omega$ is the distance between mass points, ω_0 is the angular velocity of the orbital motion of the mass centre, L is the value of the specific moment of momentum of the motion about the mass centre of the system, ψ is the angle between the local vertical and the line connecting the mass points and F_b, F_ω are given functions.

The analysis of non-resonant modes of motion of the system in sections 3.2.2, 3.2.3 shows that in the case when $(\omega_0/k)^2 = \varepsilon \ll 1$, as first approximation in ε the oscillation frequency is constant: $\dot{b} = 0$. This fact allows to simplify the task and to assume that $r = a + b\cos\omega$, where $\dot{\omega} = k$, a, b, k are constants.

Hence, we consider as model problem the motion on a circular orbit of a pendulum with periodically varying length of the bar — the motion of an orbital pendulum [94]:

$$\dot{L} = -1.5\omega_0^2 r^2 \sin 2\psi, \quad \dot{\psi} = \frac{L}{r^2} - \omega_0,$$

$$\dot{\omega} = k, \quad r = a + b\cos\omega. \tag{3.76}$$

Replacing the variable $\mathcal{L} = \dfrac{L}{a^2}$, we obtain

$$\dot{\mathcal{L}} = -\frac{3}{2}\omega_0^2 r_a^2 \sin 2\psi, \quad \dot{\psi} = \frac{\mathcal{L}}{r_a^2} - \omega_0, \tag{3.77}$$

where $r_a = 1 + z\cos\omega$, $z = b/a$. It is visible, that contrary to the case of the mathematical pendulum the change of angular orientation of the orbital pendulum does not depend on length of the bar.

3.4.2 Qualitative analysis of the attitude motion of an orbital pendulum with oscillating length

The motion of an orbital pendulum can be divided into two modes of motion — fast ($\dot\psi \gg \omega_0$) and slow ($\dot\psi \sim \omega_0$) attitude motions. For $k \gg \omega_0$ in both modes of motion there are ranges of values z, $z \ll 1$, in which the effect of longitudinal oscillations on the attitude motion is negligibly small. I.e., according to KAM theory the trajectories of the system are divided by invariant tori and are all the time close to unperturbed trajectories — to trajectories of motion of a dumb-bell satellite with a bar of constant length. The task consists in the definition of the values z and ω_0/k, for which the conditions of Kolmogorow's theorems are fulfilled. The approximate estimation of the effect of the longitudinal oscillations on the attitude motion of an orbital pendulum can be obtained from the equation

$$\frac{d^2\varphi}{d\omega^2} + 3\frac{\omega_0^2}{k^2}\sin\varphi - 2\left(\frac{d\varphi}{d\omega} + 2\frac{\omega_0}{k}\right)\frac{z\sin\omega}{1 + z\cos\omega} = 0,$$

where $\varphi = 2\psi$. From this equation follows that in the mode of slow attitude motion of the pendulum the longitudinal oscillations become negligibly small when $z \ll \omega_0/k \ll 1$. Further cases are considered when the longitudinal oscillations render essential effects on the motion of the system, i.e., either $z \sim 1$, or $\omega_0/k \sim 1$.

Let us assume that $k \gg \omega_0$, and we consider the mode of slow attitude motion of the pendulum. In this case it is possible to consider $\mathcal{L}$ and ψ as slow variables. The equations of the first approximation obtained by averaging (3.77) over ω, look like:

$$\dot{\mathcal{L}}_1 = -\frac{3}{2}\omega_0^2 s\sin 2\psi_1, \quad s = 1 + \frac{z^2}{2};$$

$$\dot\psi_1 = \frac{\mathcal{L}_1}{p_a} - \omega_0, \quad p_a = (1 - z^2)^{3/2}. \tag{3.78}$$

The equations (3.78) can be integrated in elliptical functions. The energy integral has the form

$$h = \dot\psi_1^2 + 3\omega_0^2\beta\sin^2\psi_1, \quad \beta = \frac{s}{p_a}. \tag{3.79}$$

For oscillating motions of the pendulum, $h < 3\omega_0^2\beta$

$$\sin\psi_1 = k_1 sn(\omega_0\sqrt{3\beta}(t - t_0) + F(\varphi_0, k_1^2)k_1^2), \tag{3.80}$$

where $k_1 = \max|\sin\psi_1|$,

$$F(\varphi_0, k_1^2) = \int_{\varphi_0}^{\varphi} \frac{d\varphi}{(1 - k_1^2\sin^2\varphi)^{1/2}}, \quad \sin\varphi_0 = \sin\psi_0/k_1,$$

t_0, ψ_0 is the initial time and the initial value of ψ.

The averaged oscillation period of the pendulum is

$$T = \frac{4}{\omega_0 \sqrt{3\beta}} \int_0^{\pi/2} \frac{d\varphi}{\sqrt{1 - k_1^2 \sin^2 \varphi}} = \frac{4}{\omega_0 \sqrt{3\beta}} K(k_1). \qquad (3.81)$$

Therefore, with increasing amplitude of the longitudinal oscillations the period of angular oscillations decreases, i.e., their frequency grows.

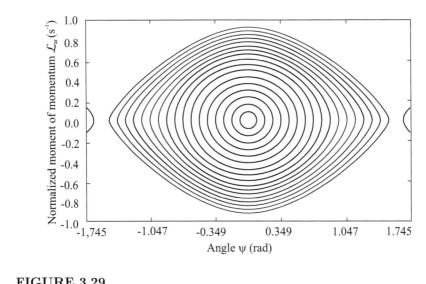

FIGURE 3.29
Phase portrait of an averaged pendulum.

The equations of first approximation (3.78) describe well the oscillations of a pendulum at rather large frequencies of longitudinal oscillations. In Fig. 3.29, Fig. 3.30 the phase portraits of the oscillations of the pendulum are shown: Fig. 3.29 — system (3.78), Fig. 3.30 — section of the phase space of the system (3.77) by the plane $r_a = 1 + z$; $k = 0.1 \, \text{s}^{-1}$, $z = 0.8$. Here and later on $\omega_0 = 0.001 \, \text{s}^{-1}$. The phase space of the system (3.77) can be considered as four-dimensional space of the variables $\mathcal{L}$, ψ, r_a, $\dot{r}_a$. On the abscissa axis the angle ψ is given in radians, on the ordinate axis in Fig. 3.29 — $\mathcal{L}_{ar} = (\mathcal{L}_1 - p_a \omega_0) \cdot 1000$ and in Fig. 3.30 — $\mathcal{L}_r = (\mathcal{L} - p_a \omega_0) \cdot 1000$ is shown. Differences in the phase portraits are seen only in areas, which are close to a separatrix separating rotatory and oscillating motions in the averaged system (3.78). A more detailed research of these trajectories shows that a slow exchange of energy (amplitude) of pendulous oscillations with transitions in rotatory and oscillating motions is seen.

The band of chaotic trajectories near a separatrix is clearly seen for $z = 0.8$ at values k smaller $0.06 \, \text{s}^{-1}$. Hence, in Fig. 3.31 a band of cross section points

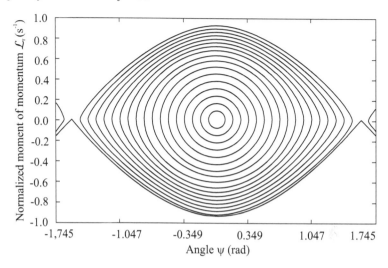

FIGURE 3.30
Phase portrait of orbital pendulum oscillations in section $r_a = z + 1$.

of the phase flow trajectory with the plane $r_a = 1 + z$ is clearly visible. The way these points are distributed allows to characterize this trajectory as chaotic. With decreasing frequency of the longitudinal oscillations keeping their amplitude fixed the bandwidth of chaotic motions grows. For other values of the amplitude of the longitudinal oscillations the chaotic motion appears at other frequencies of the longitudinal oscillations. The approximate computation of these values can be made on the basis of the equality of the relations of the frequency of pendulous oscillations of the averaged equations (3.78) to the frequency of longitudinal oscillations. In Fig. 3.32 the phase portrait of the Poincaré section is shown for $z = 0.2$ and the obtained value $k \approx 0.0191\,\mathrm{s}^{-1}$ are calculated from $z = 0.8$, $k = 0.06\,s^{-1}$.

Thus, at certain parameters of the system (amplitude and frequency of longitudinal oscillations) in the motion of the system there are trajectories, which in their evolution can be characterized as chaotic. In Fig. 3.33 the angular variation ψ in time in the chaotic regime is presented.

On the basis of the conducted analysis it is possible to assume that a source of stochastic motions is the separatrix dividing oscillatory and rotatory motions of the pendulum. Based on this supposition it is possible to give satisfactory explanations both for irregularities of trajectories and their random character, as in the unperturbed motion the separatrix corresponds to the unstable final equilibrium positions just as "the edge of a coin" divides two qualitatively different modes of motions. The approximate analysis of the dynamics of systems near the separatrix and its decomposition is a complex problem even in simpler cases [67]. At large amplitudes of the longitudinal oscillations an essential adaptation of the schemes of analysis is required. At

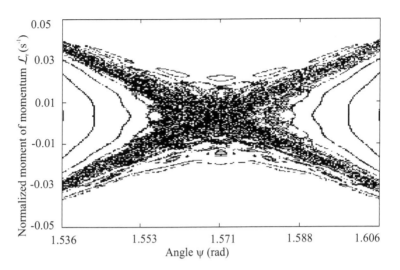

FIGURE 3.31
Phase portrait at $r_a = 1 + z, z = 0.8, k = 0.05\,\text{s}^{-1}$.

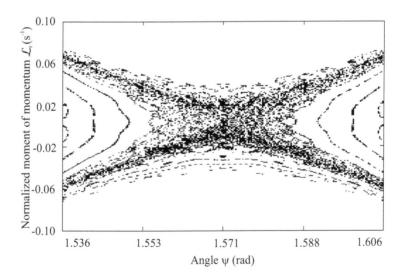

FIGURE 3.32
Phase portrait at $r_a = 1 + z, z = 0.2, k = 0.0191\,\text{s}^{-1}$.

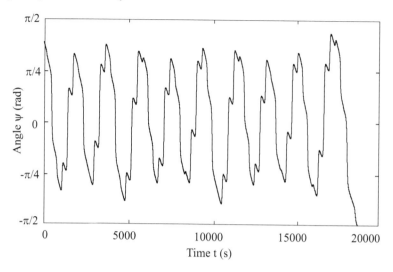

FIGURE 3.33
Variation of an angle ψ in time at $z = 0.8$, $k = 0.02\,\text{s}^{-1}$.

the same time the results of such an analysis give only approximate estimation of the bandwidth of non-regular trajectories, but do not give explanations of the phenomenon of their stochasticity. The attempt of such an explanation on the basis of the change of the topology of the separatrix results in complex mathematical problems. On the other hand [92], the numerical analysis of the system

$$\dot{\mathcal{L}} = \begin{cases} -\dfrac{3}{2} r_a^2 \sin 2\psi \,, \ \forall \, |\psi| \le 1.5359\,rad, \\[2ex] -\dfrac{3}{2} r_a^2 \operatorname{sign} \psi \sin 1.5359 \,, \ \forall \, |\psi| > 1.5359\,rad, \end{cases}$$

$$\dot{\psi} = \frac{\mathcal{L}}{r_a^2} - \omega_0, \quad r_a = 1 + z\cos\omega, \quad \dot{\omega} = k,$$

$$(3.82)$$

which is close to the original, but in which due to the absence of disturbances there are no qualitatively different motions, and, accordingly, there is no separatrix, shows that the stochastic trajectories remain, (Fig. 3.34 $(z = 0.8, \ k = 0.06\,\text{s}^{-1})$).

In an attempt of constructing the approximate solution of system (3.77) near the main resonances by an allocation procedure the supposition has appeared that a reason for the origin of chaotic trajectories are "induced resonances." The term "induced resonance" means the following. We assume it to be possible to present ψ by the way $\psi = \psi_s + \psi_k$, where ψ_s and ψ_k are accordingly slow and fast oscillating components. Then at ψ_s close to null ψ_k is "resonant" with r_a^2, and the swapping of energy in pendulous motions apparently has non-regular character. Here it is essential that the amplitude

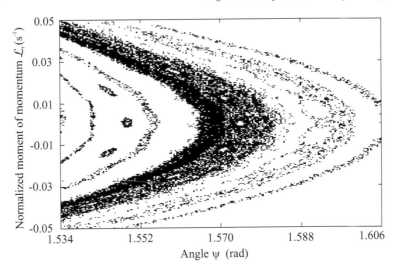

FIGURE 3.34
Phase portrait of the system (3.82) $r_a = 1 + z$, $z = 0.8$, $k = 0.05\,\mathrm{s}^{-1}$.

of the high frequency oscillations in some instants considerably surpasses the absolute value of the amplitude of the slow oscillations. Some examples of ordinary differential equations of the first order verifying this supposition were constructed. In Figs 3.35–3.40 for a number of examples the time histories and sections of Poincaré of the solution of appropriate equations are shown for zero initial conditions $t_0 = 0$, $x_0 = 0$. The Poincaré sections were plotted as follows: points in the phase plane were obtained after time intervals $\Delta t = 2\pi$.

If for the equation

$$\dot{x} = \cos(t)\cos(20\sin 0.037t + 15\cos(t)) \tag{3.83}$$

(Fig. 3.35, 3.36) with constant amplitudes and for the equation

$$\dot{x} = \cos(t)\cos(20\sin 0.037t + (0.1x + 15)\cos(t)), \tag{3.84}$$

(Fig. 3.37, 3.38) it is possible to find out periodic character of a trajectory, for the equation

$$\dot{x} = \cos(t)\cos(20\sin 0.037t + (1 + 0.1t)\cos(t)), \tag{3.85}$$

(Fig. 3.39, 3.40) with increasing amplitude of the fast oscillations the trajectory is very similar to a chaotic one. The Poincaré sections of the solution of equations (3.83) and (3.84) contain only a finite number of points. The whole band of phase points is characteristic for the equation (3.85).

The detailed research of the considered supposition is limited due to the difficult problem of the investigation of the properties of integrals of the type $\int \cos t \cos(c_1 \sin w_1 t + c_2 \cos w_2 t)dt$.

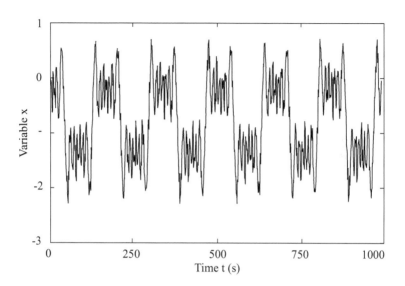

FIGURE 3.35

Trajectory of the system (3.83).

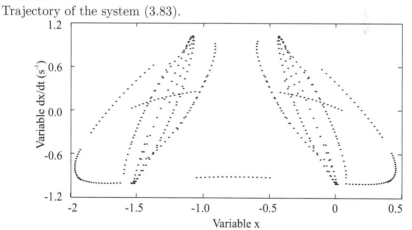

FIGURE 3.36

Poincaré section for a solution of the system (3.83).

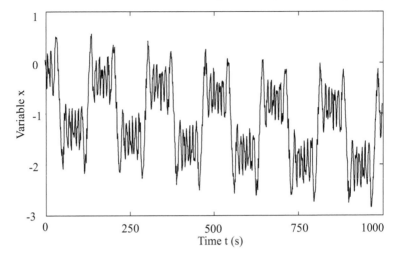

FIGURE 3.37
Trajectories of the system (3.84).

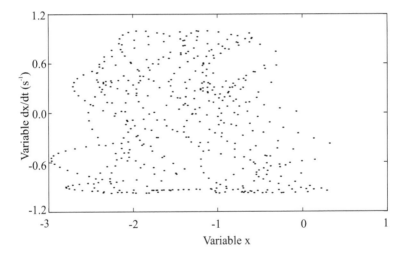

FIGURE 3.38
Poincaré section for a solution of the system (3.84).

On the other hand, it is possible to assume that the induced resonances are a main reason of the chaotic motions of the system (3.77). Then it is necessary to expect the development of the dependency of the irregularity of trajectories on the absolute value $\mathcal{L}$. However, this is not the case. Moreover,

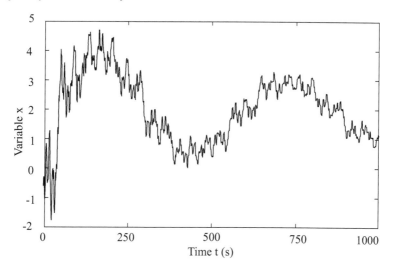

FIGURE 3.39
Trajectories of the system (3.85).

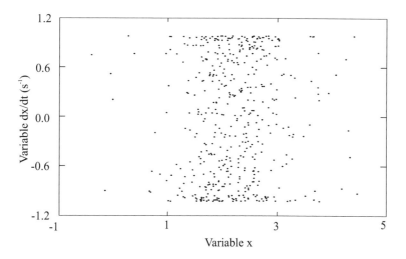

FIGURE 3.40
Poincaré section for a solution of the system (3.85).

it is necessary to expect, that in the system

$$\dot{\mathcal{L}} = -\frac{3}{2}\omega_0^2 r_a^2 \sin 2\psi, \quad \dot{\psi} = \frac{\mathcal{L}}{r_a^2} \tag{3.86}$$

the bands of chaotic trajectories should change, since in the first approximation the oscillating motion ψ (its slow motion) does not change, but $\mathcal{L}$ and, therefore, the amplitude of the fast oscillations ψ decreases for $\dot\psi < 0$ and increases for $\dot\psi > 0$. However, comparison of the phase portraits of the systems (3.77) and (3.86) both for oscillatory, and for rotatory motions of the pendulum does not give a reliable basis for the confirmation of the supposition.

The rather rough construction of the equations of the second approximation by the method of averaging gives a system of comparison of the form:

$$\dot{\mathcal{L}}^* = -\frac{3}{2}\omega_0^2 s \sin 2\psi^* + \frac{\omega_0^3}{k}\frac{\mathcal{L}^*}{p_a}I_2(\psi_0^*, t_0)\cos 2\psi^*,$$

$$\dot\psi^* = \frac{\mathcal{L}^*}{p_a} - \omega_0 - \frac{\omega_0^3}{k}\beta I_1(\psi_0^*, t_0)\sin 2\psi^*. \tag{3.87}$$

The system (3.87) correctly reflects the asymmetry of the phase space depending on the initial conditions. In Figs 3.41, 3.42 the phase portraits of the system (3.77) are represented by the section by the planes $r_a = 1$, $\dot r_a > 0$ and $r_a = 1$, $\dot r_a < 0$ respectively. This asymmetry of the phase space is easy to understand from mechanical reasons. It becomes obvious, by writing an equation for the variable $h = \dot\psi^2 + 3\omega_0^2\sin^2\psi$, reflecting the change of the energy of the pendulous oscillations

$$\dot h = -4\frac{\dot r_a}{r_a}\left(\dot\psi + \omega_0\right)\dot\psi. \tag{3.88}$$

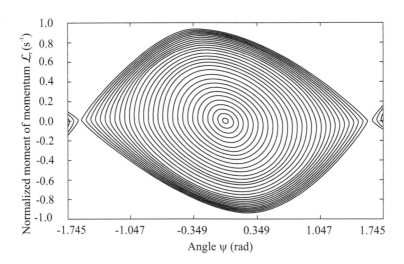

FIGURE 3.41
Phase portrait in section by the plane $r_a = 1, dr_a/dt > 0$.

The analysis of methods of research of chaotic trajectories allows to assume that any attempts of construction of approximate solutions for chaotic modes of motion are limited by unsolvable mathematical problems. More preferably

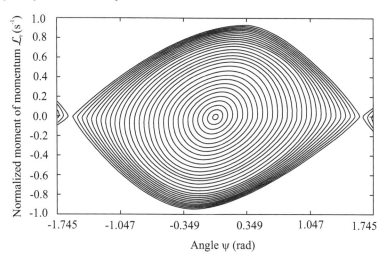

FIGURE 3.42
Phase portrait in section by the plane $r_a = 1, dr_a/dt < 0$.

looks the analysis of the transition to stochastic motions based on estimations of time discrete sequences of system states. The method of Poincaré section of the phase space relates also to those. In such approaches to the analysis, the problem of construction of the general solution of a differential equation is substituted by the research of a sequence of definite integrals of the type $\int_{t_i}^{t_{i+1}} \dot{x} dt$, where $t_i = t_0 + i*(nT_\omega)$, $T_\omega = 2\pi/k$ is the period of the longitudinal oscillations of the system, x is a variable describing the motion of the system and $i = 1, 2..., n$ are positive integer numbers. In Poincaré's method $n = 1$. Smoother sequences can be obtained for $n \geq 1$, by selecting its values for different initial conditions.

3.4.3   Analysis of a specific trajectory

Let us fix parameters of the system: $k = 0.02\,\text{s}^{-1}$, $z = 0.8$ and consider a specific regular trajectory in Fig. 3.43 [95]. Its initial conditions are equal to $\psi_0 = \pi/15\,\text{rad}, \mathcal{L}_0 = \omega_0 p_a$. In Fig. 3.43 the variable ψ is shown against the dimensionless "time" t_r, with the equal number of periods of the longitudinal oscillations: $t_r = t/T_\omega$. We consider $h^* = \mathcal{L}^2/r_a^2 + 3\omega_0^2 r_a^2 \sin^2 \psi$ as a value describing the energy of pendulous motions. Here the first item is equal to the doubled reduced kinetic energy of the relative motion, perpendicular to the bar, and the second one is the doubled reduced potential energy of the effect of the gravitational field on the relative motion of the system. The graph of change of h^* for the considered trajectory is represented in Figs 3.44, 3.45. Generally speaking, the dependence of h^* on time definitively is not simpler than the dependence of ψ on time. The analysis of the variation of h^* (Fig. 3.45) shows its essential dependence on the length of the bar.

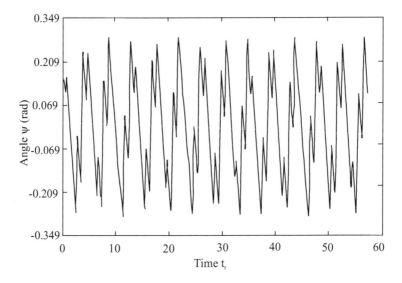

FIGURE 3.43
Regular trajectory $\psi_0 = \pi/15\,\mathrm{rad}$, $\mathcal{L}_0 = \omega_0 p_a$.

3.4.3.1 Estimation of the variation of energy for pendulous motions

The analysis of regular features of motion based on a numerical solution of the equations of motion, supposes the construction of estimates, distinctly reflecting changes in motion. The calculations display that the graph of change of h^* does not directly allow to judge about the evolution of the variations of energy of pendulous oscillations. The fact is that regular features of the variation of energy of motion on longer time intervals, such as on intervals when it is increasing or decreasing, its maximum and minimum values are discovered by local changes on each cycle of the longitudinal and pendulous oscillations, which considerably exceeds systematic changes of the energy, interesting us. In the considered case it obviously is possible and expedient to conduct the analysis of regular features of motion on the basis of the construction of a sequence of integral estimations which are discrete in time.

Following Poincaré's method we consider the sequence of values h^*, computed on each period of the longitudinal oscillations for fixed values r_a and $\dot{r}_a$. In Fig. 3.46 the sequences of values h^* for each period of the longitudinal oscillations for different fixed values of the length of the bar and for the same initial conditions are represented. As it is visible in the figure, the smoothest sequence of values h^* is obtained for a bar of fixed length, equal $1 + z$. This fact is easy to understand from equation (3.88), and also, if one considers

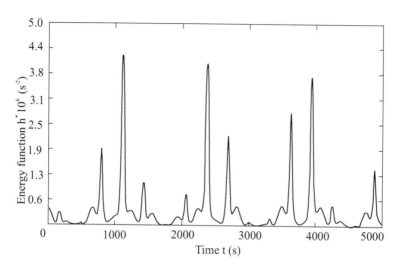

FIGURE 3.44

Variations of h^*.

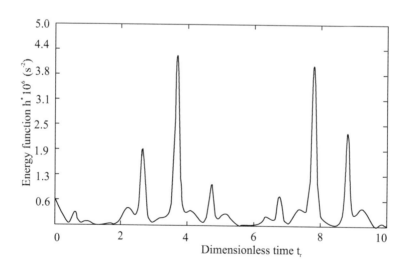

FIGURE 3.45

Variations of h^*.

trajectories of motion of the pendulum in the plane Oxy (Fig. 3.47) where $x = r_a \sin \psi$, $y = r_a \cos \psi$. In Fig. 3.47 it is visible that for the length of the bar, smaller than a certain value, the character of the pendulous motion becomes essentially more complicated and there is also the capability of inverse motions.

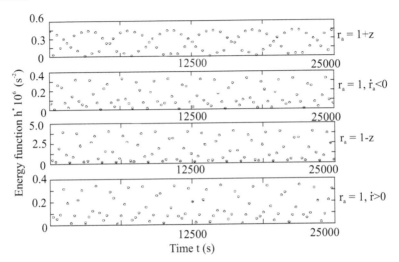

FIGURE 3.46
Sequence of values h^* in T_ω for different values of a bar length.

By reducing the time scale (Fig. 3.48), we see that in all domains of h_{1+z} changes of the same type of regular features take place. Before we begin the analysis of these regular features, we consider other ways of construction of sequences of estimations of the change of energy of pendulous oscillations.

As a measure of the change of energy of pendulous oscillations it is possible to consider the mean value h^* on each period of longitudinal oscillations

$$h_{av} = \frac{1}{T_\omega} \int_{(n-1)T_\omega}^{nT_\omega} h^* dt. \tag{3.89}$$

In Fig. 3.49 the sequence of values h_{av} obtained from (3.89) for each period of the longitudinal oscillations on the time interval equal $1000\, T_\omega$ is shown. An area of large values of h_{av} is shown in Fig. 3.50. Within some differences the character of variations in the sequence h_{av} corresponds to the variation in the sequence h_{1+z}. Analysing the possibility to use this estimated sequence, it is possible to assume that for transition of a regular trajectory into a chaotic one the deciding role should be played by the peak values of the energy of pendulous oscillations. From this point of view, the use of the sequence h_{av} does not correspond to the purpose of the research.

The construction of estimating sequences of the change of energy of pendulous oscillations is obviously possible from conditions of fixing the values

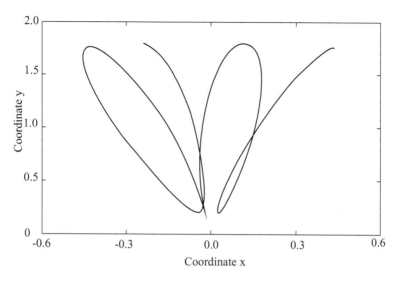

FIGURE 3.47
Trajectories of pendulum motion in a plane.

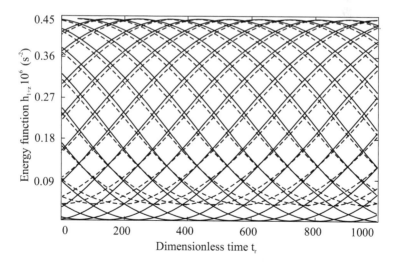

FIGURE 3.48
Sequence h_{1+z}.

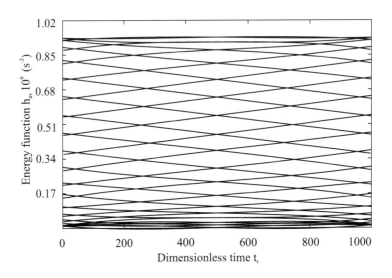

FIGURE 3.49
Sequence h_{av}.

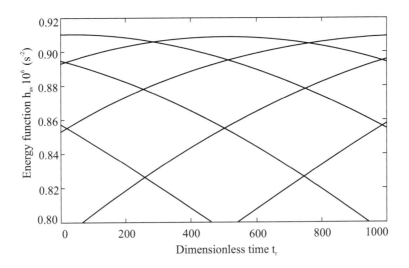

FIGURE 3.50
Sequence h_{av}.

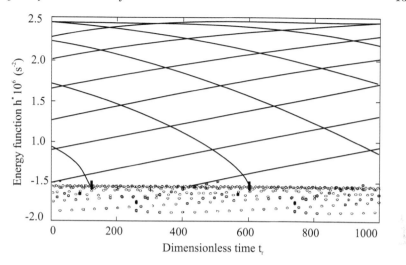

FIGURE 3.51
Sequence h^* at fixed $\psi = \pi/15\,\mathrm{rad}$.

ψ or $\mathcal{L}$ too. In Fig. 3.51 the values h^* are computed in the instants when $0.2091\,\mathrm{rad} < \psi < 0.2098\,\mathrm{rad}$. In Fig. 3.52 values h^* are shown, computed at instants for which $0.2091\,\mathrm{rad} < \psi < 0.2098\,\mathrm{rad}$ and $\dot{\psi} < 0$. From the figures, it is visible that the general regular feature of the change of sequences of values h^* has not changed. At the same time, the construction of such estimations of the change of energy of pendulous oscillations is connected to certain computing difficulties, and as it is visible from Figs 3.51, 3.52 the sequences contain "additional" values.

Considering the concept of energy of pendulous oscillations as the qualitative characteristic feature, it is possible to offer a construction of its estimating sequence on the basis of measurements of maximum values of the angle of deviation of the pendulum from the local vertical. In Fig. 3.53 the angular variation ψ for its maximum values is shown. If we draw appropriate envelope lines, it is possible to see the same general regularity in the change of the values of the estimates.

It is easy to see that the quantity of possible estimating sequences of the energy of pendulous oscillations is not limited by the listed methods. If there are general regularities of changes these sequences also have distinctions. Based on general reasoning, we consider the sequence h_{1+z} as the main estimating sequence, because it is the most simple to construct and the most appropriate according to Poincaré's method. Also we consider the sequence of maximum values of the angle ψ as most appropriate with respect to experimental research.

The offered estimating sequences do not directly supply a visual picture of

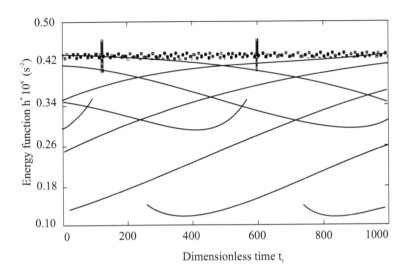

FIGURE 3.52
Sequence h^* at fixed values $\psi = \pi/15\,\text{rad}$ and $dy/dt < 0$.

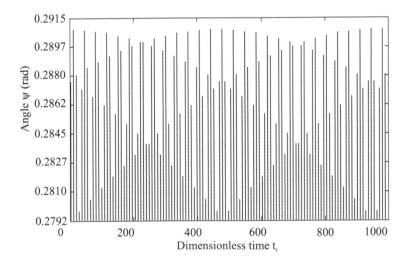

FIGURE 3.53
Variations of angle ψ maximal values.

the evolution of changes of the energy of pendulous oscillations. The envelope of the maximum values of these sequences, however, gives such a picture.

3.4.3.2 Analysis of the character of the trajectory

In all constructed estimating sequences the general regularity of their changes is seen. This regularity is connected to the almost-periodic character of the trajectory. It corresponds to the commensurability of quantities of pendulous and longitudinal oscillations in the ratio 8:35. From a more careful study of the graphs the recurrence of changes ψ and h_{1+z} in time, equal $13T_\omega$ is seen which corresponds to the commensurability of pendulous and longitudinal oscillations with the ratio 3:13.

Thus, in one specific trajectory the different commensurability of pendulous and longitudinal oscillations are exhibited. In this respect the following questions arise: if in one trajectory various commensurability are exhibited, to which commensurability is the trajectory correctly related? Is it possible to relate the trajectory to a unique resonant ratio of frequencies? Above we have defined the exhibited commensurability for a trajectory on the basis of some recurrences of changes of ψ and h^* during certain time intervals, as multiples of T_ω. This we have confirmed by constructing smooth sequences of the values of ψ and h^* during time intervals, which were multiples of the characteristic period. Now let us consider an almost-periodic trajectory with the characteristic period T_t. Then, the change of value of ψ for T_t is small, and the sequence of values ψ_i, computed at instants iT_t $(i = 1, 2, 3...)$ is smooth. By virtue of the almost-periodic character of the trajectory the longitudinal oscillations slowly transform into pendulous oscillations with the characteristic period of the trajectory. Therefore the energy transmitted into pendulous oscillations should smoothly change from period to period. Thus, the smoothness of the change of the values of ψ and the energy of pendulous oscillations in n periods of the longitudinal oscillations is one of the criteria of assessment whether the trajectory is almost periodic to nT_ω. Obviously this criterion does not answer the formulated questions.

The smooth variation of the sequences of values of ψ and h^*, are stipulated on the one hand, by the displacement of pendulous oscillations about the longitudinal ones in a characteristic period of the trajectory, and on the other hand, by the change of energy of the pendulous oscillations. Taking into account the non-linear character of the pendulous oscillations by increasing (decreasing) their energy (amplitude) as their period grows (decreases), it is possible to expect that for each trajectory there is such number n that the range of changes of values of the energy of pendulous oscillations computed through time intervals nT_ω is less than the full range of values computed during time intervals T_ω. Thus, the limited nature of the domain of values of the pendulous oscillations energy computed in time intervals nT_ω, in comparison with the full range of its values computed in T_ω, gives the second criterion

nT_ω of almost periodicity of a trajectory. Among the numbers satisfying both criteria, it is necessary to select the smallest.

The analysis of the considered trajectory ($\psi_0 = \pi/15\,\mathrm{rad}$, $\mathcal{L}_0 = \omega_0 p_a$) shows that it is almost-periodic with period equal $477 \cdot T_\omega$.

Thus, on the basis of the analysis of a specific regular trajectory it is shown that it can be related to a certain resonance of pendulous and longitudinal oscillations, and be considered as an almost-periodic trajectory. As a measure of changes of energy of pendulous oscillations it is possible to accept the sequence of maximum values of the sequence h_{1+z}, or the envelope sequence of maximum values of the angle of deviation of the pendulum from the local vertical. These measures of changes of energy of pendulous oscillations agree with the almost-periodic character of the trajectory.

3.4.4   Analysis of sets of trajectories

Let us consider a set of trajectories. In the domain of initial values ($t = 0$) we form the straight line $r_{a0} = 1 + z$, $\dot{r}_{a0} = 0$, $\mathcal{L}_0 = \omega_0 p_a$. Let us consider the set of trajectories (set 1), coming from this straight line, i.e., the set generated by changes of initial values of the angle ψ_0, the deviations of the pendulum from the local vertical. In Fig. 3.54 the phase portrait of this set is obtained by a section of the trajectories in phase space by the plane $r_a = 1 + z$ (further, if it is not stipulated especially, all phase portraits are created similarly, and for brevity we name them simply as "phase portraits"). In Fig. 3.54 the phase portrait is constructed for the variation of ψ_0 from $0\,\mathrm{rad}$ up to $1.2217\,\mathrm{rad}$ with a step-size $0.01745\,\mathrm{rad}$. For each trajectory 1000 points were plotted in the phase portrait. In the figure the trajectories are shown only for values of ψ from $-1.2217\,\mathrm{rad}$ up to $1.2217\,\mathrm{rad}$ disregarding the appearance of chaotic trajectories in the considered area. As it is visible in the figure this set of trajectories does not cover the whole space of trajectories. Therefore, besides this set, we consider similar sets of trajectories, changing the fixed initial value $\mathcal{L}_0$ (sets induced by parallel straight lines). In Fig. 3.55, similar to Fig. 3.54, the phase portrait for a set of trajectories with the initial value $\mathcal{L}_0 = \omega_0(p_a - 0.15)$ is presented.

Let us now consider more detailed the phase portraits close to the above considered trajectory, which is determined by the initial conditions $\mathcal{L}_0 = \omega_0 p_a$, $\psi_0 = \pi/15\,\mathrm{rad}$, $r_{a0} = 1 + z$. In Fig. 3.56 the phase portrait of set 1 is shown for the variation of ψ_0 from $0.2070\,\mathrm{rad}$ with stepsize $0.00035\,rad$. In the figure it is visible that the trajectory with $\psi_0 = \pi/15\,\mathrm{rad}$ really does not belong to the resonance 8:35, and relates to a resonance of higher order. The order of resonance is determined by the characteristic period of trajectories, inherent to it, $T_t = n \cdot T_\omega$: the larger the number n, the larger the order of resonance. From the figure it is also visible that the whole phase space in the considered domain is divided into resonances of high order. The resonance 8:35, as resonance of lower order in the domain is specially selected. Let us consider this case in more detail.

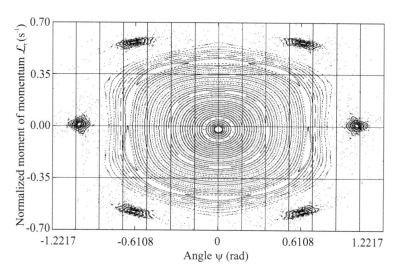

FIGURE 3.54
Phase portrait of the set 1.

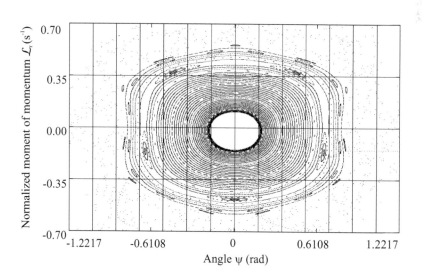

FIGURE 3.55
Phase portrait of a set $\mathcal{L}_0 = \omega_0(p_a - 0.15)$.

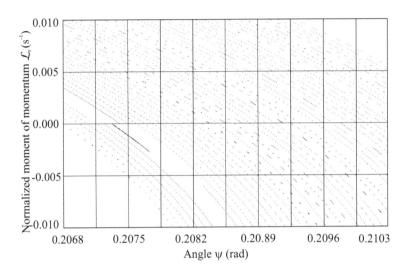

FIGURE 3.56
Phase portrait of the set 1 in domain of trajectory $\psi = \pi/15\,\text{rad}$.

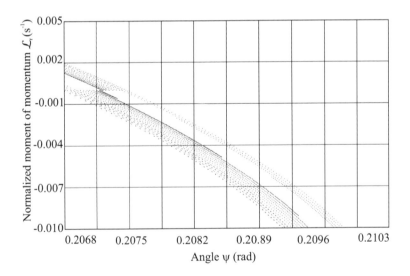

FIGURE 3.57
Phase portrait of a set in domain of resonance 8:35.

In Fig. 3.57 the phase portrait of family 1 is shown where ψ_0 changes from 0.2072 rad up to 0.2075 rad with step size 0.00001745 rad. In the figure it is visible that this resonance has the character of an almost simple commensurability of disconnected motions. Let us explain this. First, we do not observe the closed loop, which is characteristic for a non-linear resonance. Secondly, we observe a monotone increase of the density of points of the trajectory in the phase portrait if the trajectories are approaching the resonance. This is because by approaching a certain commensurability of low order of among themselves independent oscillations, the order of their commensurability grows. Approaching the resonance a certain figure formed by points of adjacent trajectories is seen. Smooth lines envelop resonances. This is connected to the increase of the period of pendulous oscillations on the considered families of trajectories with the increase of the initial value ψ_0 and with the smooth variation of the commensurability of pendulous and longitudinal oscillations. In fact, in Fig. 3.58 the phase portrait of the family $\mathcal{L}_0 = \omega_0(p_a - 0.0019)$ is presented where ψ_0 changes from null up to 0.2618 rad with step size 0.0001745. 150 points are depicted on each trajectory. Here components of the above considered figures are clearly visible.

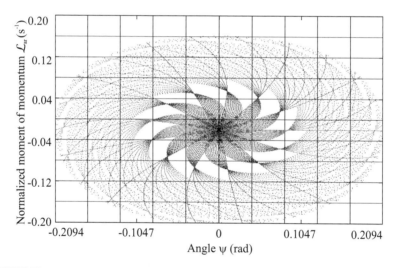

FIGURE 3.58
Phase portrait of a set $\mathcal{L}_0 = \omega_0(p_a - 0.019)$. 150 points on trajectory.

For confirmation of what was said above we consider the phase portrait of the averaged system of the equations (3.78) (Fig. 3.59). This portrait is constructed in the same way as the phase portrait shown in Fig. 3.54, i.e., the points of trajectories are created through a constant time interval, equal to the phase of longitudinal oscillations of the origin system. In Fig. 3.59 figures similar to the above mentioned are shown. In Fig. 3.60 the phase portrait of the averaged system of equations constructed similarly to the phase portrait

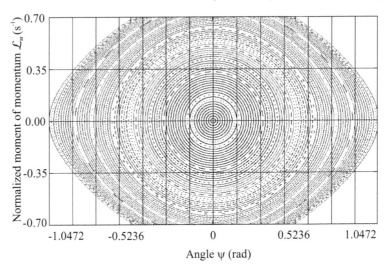

FIGURE 3.59
Phase portrait of an averaged system.

of the origin system, shown in Fig. 3.58, is presented. In Fig. 3.61 the phase portrait of the averaged system constructed in the same way as the phase portrait in Fig. 3.60 is shown, but at $z = 0.82$, approximately calculated from the condition that the period of pendulous oscillations of the averaged system T (3.78) relates to phase of longitudinal oscillations T_ω for the angle 0.1047 rad as 3:13. The presented figures confirm that for a small initial angle of pendulous oscillations the resonances have the character of an almost simple commensurability of motions disconnected among themselves.

At the same time, the considered resonance 8:35 is the resonance of a nonlinear system. The more in-depth study of the trajectories of set 1 near to this resonance has allowed to reveal the characteristic closed loop (Fig. 3.62).

Let us consider the change of energy of pendulous oscillations near resonance. In Fig. 3.63 the sequences of maximum values of the sequence h_{1+z} of set 1 for $\psi_0 = 0.2068 \, \text{rad} + i \cdot 0.00008727 \, \text{rad}$ are exhibited. Taking into account certain commensurability of trajectories each point of the sequence was plotted by selection of the maximum of the values h_{1+z} during time $35 \cdot T_\omega$. In Fig. 3.63 it is visible that on an initial interval of motion for trajectories lying below the resonance an increase of the energy of pendulous oscillations is typical, and for trajectories lying above, a decrease. In Fig. 3.64 the estimations of the change of energy of pendulous oscillations based on the evaluation of maximum values of the angle of deviation of the pendulum from the local vertical are presented. Here we observe the same regularity of the change of energy of pendulous oscillations as by passing through a resonance. The same regularity is also exhibited for other resonances of rather low order.

Thus, the analysis of trajectories for small angles of deviation of the pen-

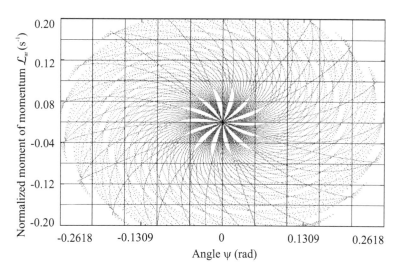

FIGURE 3.60

Phase portrait of an averaged system.

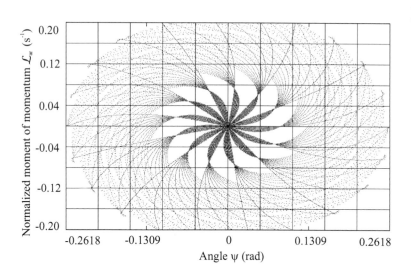

FIGURE 3.61

Phase portrait of an averaged system, $z = 0.82$.

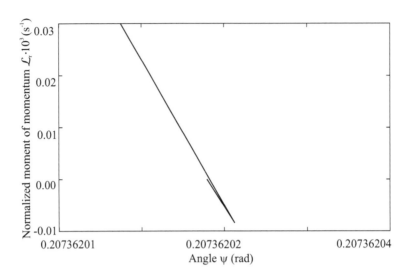

FIGURE 3.62
Loop of non-linear resonance 8:35.

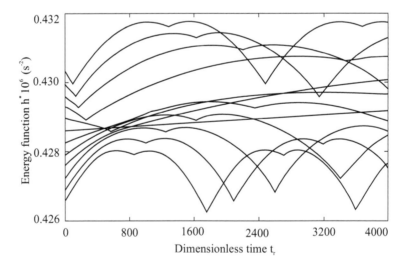

FIGURE 3.63
Envelopes of sets h_{1+z} for ψ trajectories of the set 1 with initial value $\psi_0 = 11.85 + i\,0.005$.

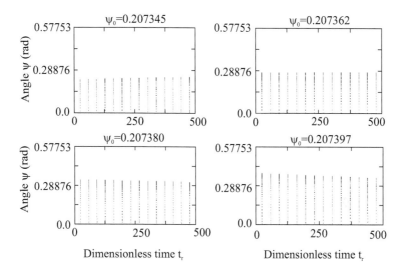

FIGURE 3.64
Variation of maximal ψ values near a resonance 8:35.

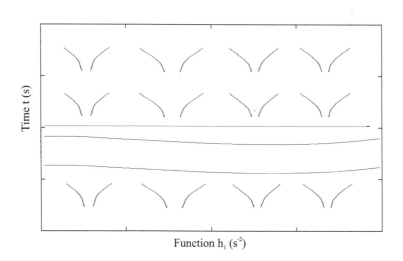

FIGURE 3.65
Representation of motion trajectories at resonance of low order as motion of ball in funnels.

dulum from the local vertical $\psi < 0.3490 - 0.4363$ rad allows to make the following conclusions [95]:

   – the whole trajectory space is divided into separate groups of resonances;
   – the non-linear character of resonances in the given area is weakly expressed, i.e., the resonant groups of trajectories are very narrow;
   – in the whole area a smooth variation of the order of commensurability of pendulous and longitudinal oscillations has taken place;
   – passing through a separate resonance of a set of trajectories induced by a separate line in the space of initial conditions, the changes of the sign of increment of the energy of pendulous oscillations has taken place. In an initial time interval, for trajectories having initially smaller values of energy, this increment is positive, and for trajectories with larger value than the value of the central resonant trajectory, it is negative.

The first conclusion poses a problem about boundaries of groups of resonances: are there trajectories separating resonances, and, accordingly, not belonging to one of the resonances? Here a numerical investigation of the considered system can not give an answer. It is only possible to assume that in the considered domain this problem is solved by the increase of the order of commensurability between longitudinal and pendulous oscillations up to infinity of the characteristic period of the trajectory.

The third conclusion states additionally that in this case a smooth variation of the character of the trajectories also takes place. In particular, in trajectories of resonances of high order the commensurability of resonances near to lower order are exhibited.

The fourth conclusion concerns, by virtue of the adopted measure of the change of energy of pendulous oscillations, resonances of rather low order only. At the same time, it is obvious that by defining the necessarily weak concept of the envelope of maximum values, we obtain also similar conclusions for resonances of high order. This conclusion links the non-linear resonance to the image of a funnel. Let us imagine the plane of the families of trajectories induced by a certain straight line in the space of initial conditions, where on the axis OY time, and on the axis OX the value of the energy of pendulous oscillations, are depicted. Then the trajectories of motion in resonances of low order can be presented as motion of a ball in appropriate funnels (Fig. 3.65), which are located along the straight line of time symmetric one by one.

The motion pattern is similar in something to motion on an infinite Halton's board, where the funnels are posed instead of nails, and for the researched regular trajectories the funnels are not overlapped, i.e., the motion of a ball happens all time in an image of the same funnel.

3.4.5   Non-linear resonances

The non-linear resonances determine the structure of the phase space of the system near the area of a stochastic layer. Resonant ratios of oscillation periods, by virtue of the increase of the period of pendulous oscillations, monotonically decrease along a straight line, generating set 1. The resonance 3:13 at $\psi_0 \approx 0.1047$ rad is the first observable resonance of rather low order, the resonance 1:6 at $\psi_0 \approx \pi/3$ is the last one.

The analysis of resonances for different parameters of the system allows to make the conclusion that their width (length of the segment of the straight line generating the set of trajectories passing through the centre of the resonance, belonging to this resonance, or the maximum increment of the angle ψ in the resonance) depends both on the order of resonance, and on the position of the resonance along the straight line generating the set 1. The order of the resonance determines the size of the reallocated energy between longitudinal and pendulous oscillations: decreasing the order of the resonance the increment of energy of pendulous oscillations grows. The arrangement of the resonance about the initial straight line of the set 1 determines the response of the system for incrementing the energy of the pendulous oscillations. It is possible to prove that approaching to $\psi_0 = \pi/2$ the width of the resonance for the same increment of energy of the pendulous oscillations is sharply increased.

For the considered parameters of the system the resonance 1:5 is the resonance of lowest order. In Fig. 3.66 for the set induced by the straight line $\mathcal{L}_0 = \omega_0(p_a - 0.162)$, passing almost through the centre of the resonance, the phase portrait of the resonant trajectories is shown. The trajectories were plotted for $\psi_0 = 0.6527 + i \cdot 0.0001745$ rad. In the figure it is visible that the phase space of the resonant trajectories is similar to the phase space of the system for small angles ψ. The whole phase space is divided into separate groups of secondary resonances — resonances between the change of the phase shift of the main resonance and the phase of longitudinal oscillations. The width of secondary resonances, similarly to the width of primary resonances, depends on the order of commensurability and their distance from the centre of the primary resonance. The width of secondary resonances depends also on the primary resonance — on its resonant ratio of oscillation periods and the arrangement along the straight line generating the set 1. The analysis shows that by decreasing the resonant ratio of phases of the primary resonance and by displacement of the centre of the resonance to a separatrix of unperturbed motion, the width of the secondary resonances increases. In Fig. 3.67 the phase portrait of the system on the boundary of the resonance 3:16 is shown. In the figure a stochastic layer is located at the boundary of the resonance, and also secondary resonances, located closely to this layer, are clearly visible.

The investigations of the change of energy of pendulous oscillations of resonant trajectories allow to link the resonance of rather low order to the image of the spatial funnel, on the surface of which the grooves corresponding to secondary resonances are marked. An image of motions in non-linear resonances presented in Fig. 3.65 changes also accordingly. Let us imagine the plane OXZ generated from straight lines, where on the axis OX the energy of pendulous oscillations, and on the axis OZ the angle ψ is drawn. Then the change of parameters of motion in resonances of low order in time can be presented as motion of a ball in appropriate spatial funnels and their grooves

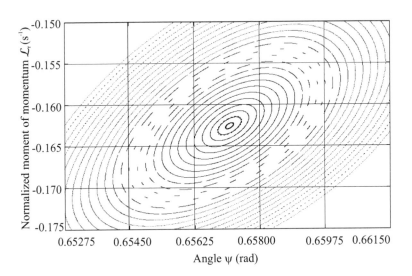

FIGURE 3.66
Phase portrait of a non-linear resonance 1:5.

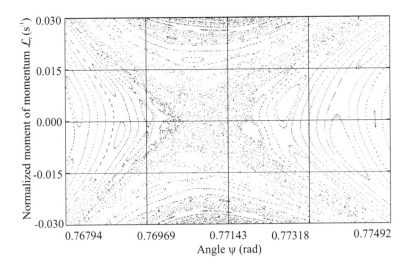

FIGURE 3.67
Phase portrait on boundary of a resonance 3:16.

located along the straight time axis OY is symmetric one by one. Actually the motion pattern in non-linear resonances is more complex, since for secondary resonances there can be again secondary resonances and so on.

3.4.6 Image of chaotic motions

The increase of the density of the arrangement of resonances of low order and width of secondary resonances for approaching a stochastic layer allows to make the conclusion or supposition, that the reason of the origin of chaotic motion is the overlap of resonances [95]. A chaotic trajectory, contrary to a regular one, does not belong to any specific commensurability of pendulous and longitudinal oscillations. It passes through a group of resonances, "jumping" in its motion on adjacent commensurability of this group.

The image constructed above of the non-linear resonance allows also to construct a visual image of a chaotic trajectory. For an appropriate initial position of a ball all overlapping funnels are accessible to its motion in overlapping funnels (the overlapping parts of the funnels need to be removed). For an initial position of the ball at the bottom of the funnel it can not get into adjacent funnels. The conformity of the constructed image to real motions of the system is confirmed by "islands" of resonances in the stochastic layer of oscillatory and rotatory motions of the pendulum in Figs 3.54, 3.55, and by the investigation of the stochastic layer only for oscillating motions of the pendulum near the boundary of the resonance 3:16 (Fig. 3.67).

3.4.7 Effect of energy dissipation

The effect of dissipative forces on chaotic modes of motions is further considered. As a model problem of the dynamics of space tether systems the model of a mathematical pendulum with periodically varying length of the bar on a circular orbit (orbital pendulum) is considered [96]. The dissipative forces are modelled as external forces of viscous friction. The derived equations of motion have the form

$$\dot{\mathcal{L}} = -\frac{3}{2}\omega_0^2 r_a^2 \sin 2\psi - \xi r_a^2(\dot{\psi} + \omega_0), \quad \dot{\psi} = \frac{\mathcal{L}}{r_a^2} - \omega_0, \qquad (3.90)$$

where $r_a = 1 + z\cos\omega$, $z = b/a$, b, a are the mean length of the bar and the amplitude of its oscillations respectively, ω_0 is the angular velocity of the orbital motion of the pendulum (similarly as before ω_0 is adopted to be equal to $0.001\,\mathrm{s^{-1}}$), ψ is the angle between the local vertical and the bar, $\dot{\omega} = k$, ξ is the factor of external viscous friction ($\xi \ll 1$) and ξ, a, b, k are constants.

3.4.8 Results of numerical calculations

Examples of typical behaviour of trajectories of a system (3.90) in chaotic modes of motion are investigated below.

In Figs 3.68–3.71, for parameters of the system $z = 0.8$, $k = 0.02\,\mathrm{s}^{-1}, \xi = 10^{-6}\,\mathrm{s}^{-1}$, the images of the phase trajectory with the same initial values $\psi_0 = 0.8727\,\mathrm{rad}$, are exhibited; $\mathcal{L}_0 = \omega_0 p_a$ at the instant $t = 0$. As before, the phase space of the system (3.90) is considered as space of four variables $\mathcal{L}$, ψ, r_a, $\dot{r}_a$. The images of the phase trajectories (their phase portraits) are created through a Poincaré map — by section of the phase space by the plane $r_a = 1 + z$, $\dot{r}_a = 0$. $\mathcal{L}_r = (\mathcal{L} - p_a\omega_0)1000$, $p_a = (1 - z^2)^{3/2}$. Calculations were conducted with the Runge–Kutta method of the 4th order with a constant stepsize of integration. Differences in the calculations of trajectories shown in Figs 3.68–3.71, result only from the selection of the integration stepsize. In Fig. 3.68 the change of the trajectory is exhibited for an integration stepsize equal $h_g = \pi/(521 \cdot k)$, in Fig. 3.69 $h_g = \pi/(621 \cdot k)$, in Fig. 3.70 — $h_g = \pi/(651 \cdot k)$, in Fig. 3.71 $h_g = \pi/(731 \cdot k)$. In each case the trajectory eventually arrives in some area of the phase space and remains there. For the trajectory shown in the Fig. 3.68, such an area is the area of the resonance 1:2 (one rotation of the pendulum for two periods of longitudinal oscillations) for a rotation opposite to the orbital one. For the trajectory shown in Fig. 3.69 it is the area of oscillation of the pendulum near an equilibrium position of the pendulum with constant length of the bar $\psi = \pi$ (further on, for brevity, we call these areas simply as areas of oscillations near an equilibrium position, though here a steady mode of oscillations takes place with commensurability 1:1 — one oscillation of the pendulum for one phase of longitudinal oscillations). For the trajectory shown in Fig. 3.70 it is the area of the secondary resonance of the 1:2 resonance of direct rotations. In Fig. 3.71 it is the area of the 1:1 resonance of direct rotations.

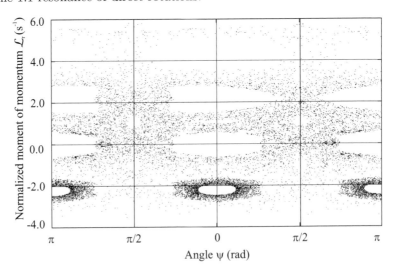

FIGURE 3.68

Image of a trajectory for $h_g = \pi/(521\,k)$.

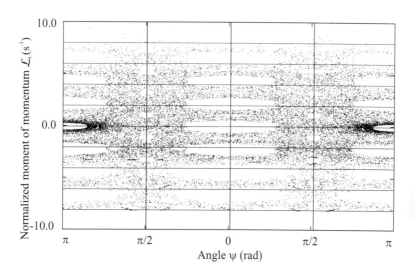

FIGURE 3.69
Image of a trajectory for $h_g = \pi/(621\,k)$.

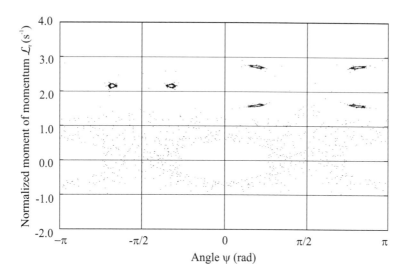

FIGURE 3.70
Image of a trajectory for $h_g = \pi/(651\,k)$.

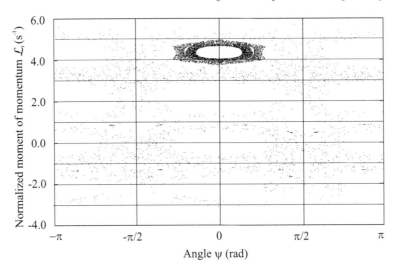

FIGURE 3.71
Image of a trajectory for $h_g = \pi/(751\,k)$.

A similar picture of change of trajectories takes place for a constant integration step for minor changes of the initial data also. In Fig. 3.72 for the parameters of the system $z = 0.2, k = 0.002\,\text{s}^{-1}$, $\xi = 10^{-5}\,\text{s}^{-1}$ the phase portraits of trajectories with the initial data $\mathcal{L}_0 = 0.006\,p_a$ and $\psi_0 = 0.0001745 \cdot i$ are exhibited where i varies from 0 up to 16. The integration stepsize for the calculation was equal $h_g = \pi/(521 \cdot k)$. Here the attracting areas for the trajectories are the areas of 1:1 resonance of inverse rotations; 3:2 resonance of direct rotations and the area of oscillations near the final equilibrium positions $\psi = 0$, $\psi = \pi$. Changes of the integration stepsize and changes of the initial data do not change noticeably the picture shown in Fig. 3.72.

In Fig. 3.73 for parameters of the system $z = 0.8$, $k = 0.01\,\text{s}^{-1}, \xi = 0.0005\,\text{s}^{-1}$ the image of the phase trajectory is exhibited with the initial data $\mathcal{L}_0 = 0.04$, $\psi_0 = 0$. Here the set of attracting trajectories has a complex structure and the motion in this area, contrary to the motion in earlier considered attracting areas, has non-regular character. This area is robust against changes of the initial conditions, integration step size and small variations of parameters of the system.

3.4.9   Analysis of chaotic motions and their images

The offered image of chaotic motions linking non-linear resonances to the image of a spatial funnel, and its secondary resonances with tucks of the surface of this funnel, allows also to explain the effect of dissipative forces on the behaviour of trajectories [96].

If dissipative forces are absent, the chaotic trajectory of motion passes

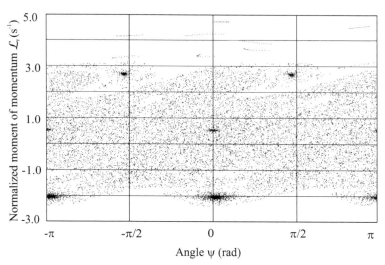

FIGURE 3.72

Phase portrait of a trajectory at $z = 0.2, k = 0.002\,\text{s}^{-1}$, $\xi = 0.00001\,\text{s}^{-1}$.

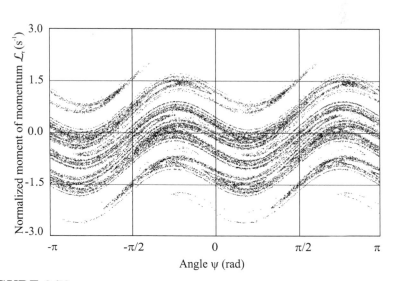

FIGURE 3.73

Phase portrait of a trajectory at $z = 0.8, k = 0.001\,\text{s}^{-1}$, $\xi = 0.0005\,\text{s}^{-1}$.

through cascades of primary, secondary, etc. resonances ("cascade of spring-boards") increasing and decreasing the energy of the attitude motion of the pendulum. The whole area of chaotic motions as an area of overlapping of non-linear resonances can also be presented as some large funnel. Generally speaking, the whole area of chaotic motions is accessible to a chaotic trajectory, and by virtue of the Poincaré's theorem about recurrence, the trajectory permanently returns close to each of its points. The elimination of this area is made only by areas of trajectories with initial data appropriate to rather fine tuning on a resonance of low-level order of longitudinal and angular motions of the pendulum (located at the bottom of the resonant funnel) and a trajectory of pendulum oscillations about an equilibrium state. The motion pattern is similar to the motion of a ball in a roulette. If friction is absent the ball permanently moves around in the wheel of the roulette, dropping in the compartments, but not remaining in them. Only for certain initial parameters of motion of the ball it can not leave a compartment.

The effect of dissipative forces is always directed to decrease the energy of attitude motions of the pendulum ($\xi \omega_0 r_a^2$ may be considered as a conservative effect). The chaotic trajectory, as well as in the case of absence of dissipative forces, passes through cascades of resonances and increases and decreases the energy of the attitude motion. However, under the action of dissipative forces it becomes possible that the trajectory hits the area of non-linear resonance (in the not overlapping part of the resonant funnel) or the field of oscillations near the final equilibrium positions.

For certain parameters of the systems this capability should be eventually also realised. Getting into a resonance, the trajectory remains there "for ever," if the resonance can supply pumping of energy of the attitude motion, equal to its dispersion by dissipative forces over the period of the motion. The stability of the motion in this case is stipulated by a local minimum of the power of the resonant trajectory. The motion pattern (Figs 3.68–3.72) is really similar to a roulette, where the ball appears in one of the compartments in a random way. Only the image of the compartment does not correspond to the observable motion. The image of funnel (funnel, similar to whirlpool) only is appropriate here, since the trajectory can be either tightened in this funnel, or thrown out from it with an essential increment of the energy.

For an increase of the effect of the dissipative forces (increase of the factor of viscous friction ξ) the quantity of resonances, which are capable of supplying appropriate swapping of energy and stability of resonant trajectories (Fig. 3.72), decreases. Thus only the resonances of low order, such as 3:2, 1:1, 1:2 remain. In Fig. 3.74 for parameters of the system $z = 0.8$, $k = 0.02\,\mathrm{s}^{-1}$, $\xi = 0.0001\,\mathrm{s}^{-1}$ the phase portraits of trajectories with initial data $\mathcal{L}_0 = 41 p_a \omega_0$ are exhibited. $\psi_0 = \pi/3 + i \cdot 0.01745\,\mathrm{rad}$, where i varies from $0.01745\,\mathrm{rad}$ up to $\pi/6$ with a stepsize $0.01745\,\mathrm{rad}$. Here the resonance 1:4 was added to the listed resonances.

The further increase of dissipative forces results in areas of oscillations near final equilibrium positions $\psi = 0$, $\psi = \pi$ that are unique attracting sets.

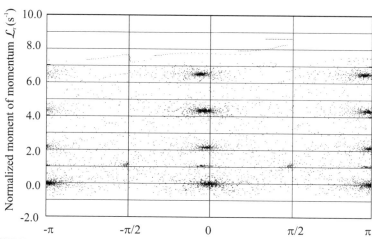

FIGURE 3.74

Phase portrait of a trajectory at $z = 0.8, k = 0.02\,\mathrm{s}^{-1}$, $\xi = 0.0001\,\mathrm{s}^{-1}$.

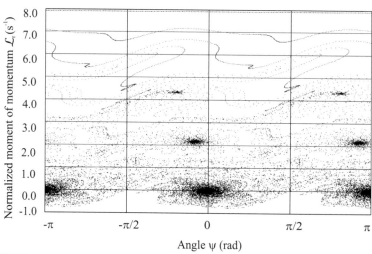

FIGURE 3.75

Phase portrait of a trajectory at $z = 0.8, k = 0.02\,\mathrm{s}^{-1}$, $\xi = 0.0004\,\mathrm{s}^{-1}$.

Hence, any trajectory of motion of the system for parameters $z = 0.8$, $k = 0.02\,\mathrm{s}^{-1}$ and at $\xi > 0.001\,\mathrm{s}^{-1}$ rather quickly reaches the area of oscillations near equilibrium positions. In Fig. 3.75 for parameters of the system $z = 0.8$, $k = 0.02\,\mathrm{s}^{-1}$ $\xi = 0.0004\,\mathrm{s}^{-1}$ the phase portraits of trajectories with the initial data $\mathcal{L}_0 = 0.0079 - p_a\omega_0$ are shown $\psi_0 = -\pi + i \cdot 0.01745\,\mathrm{rad}$, where i varies from $0.01745\,\mathrm{rad}$ up to 2π with a step size $0.01745\,\mathrm{rad}$. Here still some trajectories reach and remain in resonant funnels of resonances 1:1 and 1:2. Fig. 3.76 is constructed similarly to Fig. 3.75. Only here $\xi = 0.0009\,\mathrm{s}^{-1}$. In this case steady resonant trajectories are not present. In Figs 3.75, 3.76 changes of the character of trajectories are clearly visible for different values of the moment of momentum (angular velocity) of the motion of the pendulum. For large angular velocity in trajectories their continuous dependence on the initial data (small change of the initial data results in a small change of the trajectory, see the curves in top of figures). In approaching the attracting sets the effect of their intermixing is seen. It is possible to explain it by the fact that for rather large angular velocity the possible swapping of energy by internal forces during the period of longitudinal oscillations results in essentially less dispersion of energy by the dissipative forces. The linear dependence of dissipative effects on the angular velocity results in a decrease of dispersion of the energy for a decrease of angular velocity.

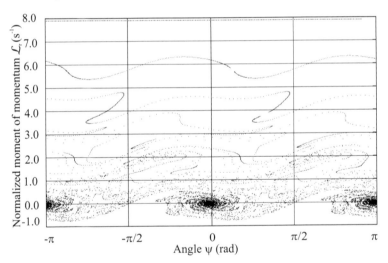

FIGURE 3.76
Phase portrait of a trajectory at $z = 0.8, k = 0.02\,\mathrm{s}^{-1}$, $\xi = 0.0009\,\mathrm{s}^{-1}$.

Remaining of the resonances also speaks about a comparability of possible pumping of the energy with their dispersion. The capability of "jumping" of a trajectory through separatrices of motions from one resonance to another remains also in this case. Here it is possible to offer the following image of motions. The effect of dissipative forces can be presented as a flow directed from outer boundaries of the chaotic funnel to the centre of this funnel — to the final equilibrium positions, and whose pressure depends linearly on the energy of rotatory motions of the pendulum — on the altitude above the centre of the funnel. For fixed parameters this flow completely determines the character of the motion of the imagined ball: the flow carries it.

For an attenuation of the force of the flow in approaching the centre of the chaotic funnel the flow performs a whirling motion above the resonant funnels (Figs 3.75, 3.76) but still completely determines the character of the motion. For the further attenuation of the flow it is not capable any more to carry the imaginary ball from the steep resonant funnel (Fig. 3.75) and is not capable "to cover" separatrices of motions near final equilibrium positions (Figs 3.75, 3.76).

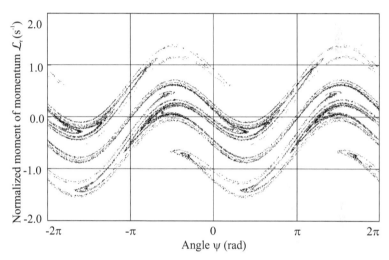

FIGURE 3.77
Variation of a strange attractor at increase of damping.

For certain parameters of the system, such as $z = 0.8$, $k = 0.01\,\text{s}^{-1}$ and $z = 0.9$, $k = 0.02\,\text{s}^{-1}$, the oscillatory modes of motions near final equilibrium positions in the absence of the dissipative forces are impossible (at any initial data the trajectory passes into a rotatory mode of motion). In this case for rather large parameter values ξ, such that there are no more stable motions in resonant funnels, so-called "strange attractors" can be the uniquely attracting set (Fig. 3.73). In the motion on a strange attractor the trajectory passes through cascades of separators of motions, sometimes reaching the maximum values of the energy, but under the action of the dissipative forces again returns in areas, close to a final equilibrium position. In Fig. 3.73 "tongues" are clearly visible where the trajectory reaches its maximum values of energy. In Fig. 3.77 the change of the strange attractor is exhibited for an increase of dissipative effects: $\xi = 0.001\,\text{s}^{-1}$ ($z = 0.8$, $k = 0.01\,\text{s}^{-1}$). For $\xi = 0.002\,\text{s}^{-1}$ and the same parameters of the system the attracting set is the resonance oscillation near the equilibrium state.

For parameters of the system $z = 0.9$, $k = 0.02\,\mathrm{s}^{-1}$ it was not possible to find out the existence of the strange attractor for different values ξ. Here, increasing ξ steady resonant motions always exist near an equilibrium state. The conducted analysis allows to assume that the existence of strange attractors is connected to the absence of steady-state trajectories in the motion of the system and to the limitation of area of motions. In this sense the construction of mathematical criteria for the existence of strange attractors is possible.

4

Use of Resonance for Motion Control

4.1 Introductory remarks: Formulation of the problem

The control of the motion of a system of rigid bodies in a central force field by internal forces keeps the total moment of momentum of the system constant. Therefore the strategy of control of such systems can consist only in the redistribution of the moment of momentum between relative motion of the system and motion of its mass centre and (or) in the change of the kinetic energy of the system at the expense of the work of internal forces. An example of control of the motion of the system for which only its kinetic energy changes is the known scheme of the gravyplane [20].

The idea of the gravyplane has been developed in [34, 53, 73, 74]. In all these works the principle of control of the orbital motion "remains the same, in a change of the orbit of the gravyplane at the expense of the variation of the force of gravitation acting on it" [20]. By variation of the forces of gravitation in all these works is meant a change of the gravitational forces which is in resonance with the orbital motion. This will be carried out by means of acting on the mass geometry of the space vehicle.

In this chapter within the framework of the model problem of connected point masses the capabilities of a number of control schemes for the motion of the systems based on the reallocation of the moment of momentum between orbital and relative motions will be studied [89]. Thus the control of the orbital motion of the rigid system (dumb-bell) is realised by the way of maintenance of the prescribed orientation of the system in the orbital system of coordinates, and by control of the orbital motion of the system with flexible connection (tether), by means of change of length of the connection in modes which both are in "internal" and "external" resonance, i.e., in modes of resonance with rotation of the tether about the centre of mass and in modes of resonance with the orbital motion.

The opportunity of control of the motion of a mechanical system with the help of internal forces in a Newtonian force field is caused by its heterogeneity, and the control efficiency essentially depends on the linear extension of the system. Space cable systems the projects of application of which assume creation of space systems with an extension of up to hundreds of kilometers can allow effectively the use of internal forces for the control of the system's motion.

We assume that the distance of the tether between the bodies changes in some given mode by the application of internal forces along the line between the connected bodies. Then the translational and rotational motion of the system is described by equations (3.55), (2.87) where the perturbing accelerations up to ε_1^2 are determined by the formulas (3.54). We connect the non-rotating system of coordinates with the constant moment of momentum of motion of the system (section 1.2.2). Then in motion of the system three first area integrals (3.58) exist.

We consider that the distance between the bodies r is a known function of the angular variables of the system (3.55), (2.87) (angles of orientation of the motion of the system) and that r changes periodically near some value a with amplitude b and can be represented as $r = a - b\cos\omega$ where the frequency of the longitudinal oscillations ω is a function of the angles of orientation of the system.

Equation (3.56) allows to determine the necessary control action directed along the line connecting the bodies (line of the tether) in fulfilment of this functional dependence. The possibility of the realisation of such a control for unilateral connections is determined by the condition

$$T = -\ddot{r} + \frac{L^2}{r^3} + F_r \geq 0.$$

4.2 Control of motion of the system around its mass centre

Let us consider possibilities of control of the relative motion of the system [91]. As the influence of the Newtonian force field on the relative motion contains terms proportional to the first degree of the value r/R, for the investigation in the first approximation we can neglect the terms containing r/R in the second and higher degree. Then the trajectory of the motion of the mass centre is the unperturbed Keplerian orbit and the perturbing accelerations of the relative motion are described by the force function (3.1).

We estimate the possibilities of change of the value of the moment of momentum of the relative motion for the motion of the system in the plane of the circular orbit [89]. The equations (3.55) have the form

$$\dot{L} = -\frac{3}{2}\frac{\mu}{R^3}r^2\sin 2\lambda, \quad \dot{\lambda} = \frac{L}{r^2} - \dot{\nu}, \quad \dot{\nu} = \sqrt{\frac{\mu}{R^3}}, \tag{4.1}$$

where $r = a \pm b\cos\omega$, and λ is the same expression as in equations (3.18).

Carrying out the formal averaging with respect to the angular variable λ and ω, it is easy to see that in the first approximation the change of L is possible only for the resonant tuning of longitudinal oscillations of the kind 1:1

and 1:2. Thus, the mean period of change of velocity L of the system rotation around the mass centre at resonance 1:1 $(\omega = \lambda + \omega_0$ where ω_0 is the initial phase of longitudinal oscillations) is proportional to the value $0.5\mu/R^3 b^2$, and at resonance 1:2 $(\omega = 2\lambda + \omega_0)$ it is proportional to the value $\mu/R^3 ab$. Since the greatest possible value b does not exceed a, and the increase of the amplitude of longitudinal oscillations requires essentially an increase of the magnitude of the control force, the use of the "swings" of resonance 1:2 is, obviously, more preferable.

For the construction of estimations of the possible increase of the moment of momentum of the relative motion of the rotating tether we consider a law of change of the length of the tether of the following kind

$$r = A - b \, sign(\sin 2\lambda). \tag{4.2}$$

It is obvious that for such a law of change of the length of the tether the increase of L is maximum.

Using the solution of the equations (4.1) for regions of constant length of the tether [20], we obtain that the change of angular velocity of the relative motion for one period of the longitudinal oscillations is described by the formulas

$$\lambda_1'^2 = \lambda_0'^2 + 3, \quad \lambda_2' = \beta^2(\lambda_1' + 1) - 1,$$

$$\lambda_3'^2 = \lambda_2'^2 - 3, \quad \lambda_4' = \frac{1}{\beta^2}(\lambda_3' + 1) - 1. \tag{4.3}$$

Here the prime designates derivative with respect to ν, $\beta = (a+b)/(a-b)$, λ_0' and λ_4' are, respectively, initial and final angular velocities, the initial value of λ_0 is assumed to be equal $\pi/2$.

Change of ν for the same time is

$$\Delta\nu = K\left(\frac{3}{\lambda_0'^2 + 3}\right)\Big/\sqrt{\lambda_0'^2 + 3} + K\left(\frac{3}{\lambda_2'^2 + 3}\right)\Big/\sqrt{\lambda_2'^2 + 3}, \tag{4.4}$$

where K is the complete elliptic integral of the first kind.

In Fig. 4.1 the dependence on the period of the mean angular acceleration of the relative rotation of the tethered system on the value of its initial angular velocity for various values of the amplitude of longitudinal oscillations is shown. The transition from initially librational motion of the tethered system about the local vertical into rotatory motion is realised by the way of swinging the tethered system. For this purpose the change of distance between the bodies can be determined as

$$R = a - b \, sign(\lambda' \sin 2\lambda). \tag{4.5}$$

In Fig. 4.2 for an initially circular orbit of motion of the mass centre of radius $R = 6671$ km and $\lambda_0' = 0$ the change of the parameters of motion is presented for the given laws of control for $b = 0.1a$, $a = 200$ km. The continuous lines show results of the numerical solution of the complete equations of

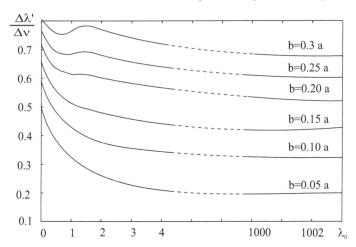

FIGURE 4.1
Mean value $\frac{\Delta\lambda'}{\Delta\nu}$ calculated over one period of longitudinal oscillation.

motion, the dotted line the solution of the equations (4.1) averaged with respect to λ, the circles denote results of computations based on formulas (4.3). For the prescribed values the motion of the tether after one swing passes into rotatory motion. As it is visible from the figure the computations of rotatory motion on the basis of the formulas (4.3) agree better with the results of numerical integration than with the solution of the averaged equations. At the same time the expected deviations of the results grow with time in view of the decrease of the radius of the orbit of the mass centre.

From what has been said above it follows that by the appropriate resonant tuning of the change of the tether length it is possible to considerably increase the angular velocity of the relative motion and consequently the relative linear velocities of the bodies too. The projects of the use of Space cable systems for launching a payload on higher orbits are based on cutting the tether (setting the bodies free) in the right moment of time.

The use of the project "swing" for the initial boost of the bodies will allow considerably to expand the capabilities and will increase the radius of the orbit of manoeuvre of the payload.

Since the restrictions on the achievable velocity of the motion of bodies about the mass centre are defined only by restrictions on the control T of the acceleration, the case is possible that the absolute velocity of one of the bodies becomes large enough for the transition to a hyperbolic trajectory. Since for a hyperbolic trajectory of motion the condition

$$V_i^2 > \frac{2\mu}{R_i} \approx \frac{2\mu}{R}, \quad i = 1, 2 \tag{4.6}$$

must be satisfied, where V_i is the velocity of the motion of the ith body about

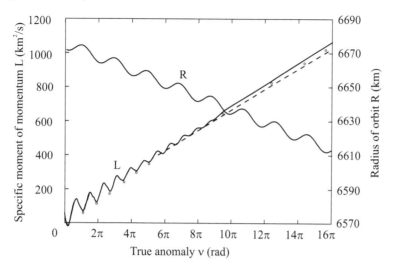

FIGURE 4.2
Variations of L and R at $r = a(1 - 0.1sign(\lambda' \sin 2\lambda))$.

the Newtonian attractive centre, for $\lambda = 0$ the relative velocity of one of the bodies should exceed the value $(\sqrt{2} - 1)\sqrt{\mu/R}$. If under these conditions the tether is cut off, one of the bodies will pass to a hyperbolic trajectory.

In table 4.1 for a circular orbit of radius $R = 6671$ km and for initial conditions $\lambda_0 = \pi/2$, $\lambda' = 10^{-5}$ for the law of control (4.2), $b = 0.1a$ estimations are presented to reach a sufficiently large velocity that for one of the bodies of a system with equal masses a transition to a hyperbolic orbit is possible.

In the fifth column of the table 4.1 the value of the centrifugal acceleration corresponding to the angular velocity is presented. For the rotation of the tethered system with bodies of different mass, the parameters for transition of

TABLE 4.1
Estimations for reaching the velocity that is
sufficient for the transition to a hyperbolic
trajectory of one of the bodies after cutting the
tether

a, km	λ'	ν, turn	*Time*, day	T, m s^{-2}
1	5526.0	2762.0	173.3	41015.0
10	552.0	276.0	17.3	4106.0
50	110.0	54.0	3.4	827.1
100	55.0	27.0	1.7	421.1
200	27.0	13.0	0.8	210.5
500	10.1	4.3	0.3	82.7

the lighter body to a hyperbolic trajectory are improved almost in $2m_2/(m_1 + m_2)$ times where m_2 is the mass of the heavier body.

The parameters of transition can also be improved for the motion on an elliptic orbit [59].

The motion of the second body after cutting off the connection depends on the distance to the mass centre. In particular, it is easy to select such mass parameters of the system that the second body remains practically on the same orbit but rotates in the opposite direction.

In fact, let $V_1^2 = 2\mu/R$, $V_0^2 = \mu/R$ where V_0 is the velocity of motion of the mass centre, then the velocity of the second body is equal to

$$V_2 = -V_0 - (\sqrt{2} - 1)\sqrt{\frac{\mu}{R}\frac{r_2}{r_1}}$$

where r_i is the distance from the mass centre to the ith body. Taking into account that $r_2 = r_1 m_1/m_2$ we obtain that for the mass ratio $m_2/m_1 = (\sqrt{2} - 1)/2$ the velocity of the second body is equal to $V_2 = V_0$. I.e., the second body obtains a velocity equal in value to the orbital velocity of the mass centre but directed into the opposite direction.

The possibility of control of the orientation of the moment of momentum of the relative motion we consider under the assumption that the angular velocity of rotation of the tethered system about its mass centre significantly exceeds the angular velocity of the orbital motion. We show that up to oscillating terms with the help of control of the length of the tether it is possible to obtain practically any value of the angles θ and ψ.

The equations (3.59) in the considered case look like

$$\frac{d\theta}{d\nu} = 0, \quad \frac{dL}{d\nu} = 0, \quad \frac{d\psi}{d\nu} = -\frac{3}{4}\sqrt{\frac{\mu}{p^3}\frac{r^2}{L}}\cos\theta. \tag{4.7}$$

Therefore, for $\theta \neq \pi/2$ for obtaining the prescribed value of ψ control is not required.

The change of the angle of nutation θ requires resonant tuning of the change of length of the tether. We consider two simple laws of change of length of the tether based on tuning of the "internal" resonance of system $r = a \pm b \sin 2\varphi$ and the "external" resonance $r = a \pm b \sin 2(\nu - \psi)$.

For the control of $r = a \pm b \sin 2\varphi$ the equations averaged over φ and ν constructed in agreement with the above stated scheme of averaging look like

$$\frac{d\theta}{d\nu} = \mp\frac{3}{4}\sqrt{\frac{\mu}{p^3}\frac{r_1^{*2}}{L}}\sin 2\theta,$$

$$\frac{d\psi}{d\nu} = -\frac{3}{4}\sqrt{\frac{\mu}{p^3}\frac{r_2^{*2}}{L}}\cos\theta, \tag{4.8}$$

$$\frac{dL}{d\nu} = \mp\frac{3}{2}\sqrt{\frac{\mu}{p^3}}r_1^{*2}\sin^2\theta,$$

where

$$r_1^{*2} = \frac{a^3 b + 0.75ab^3}{a^2 + 0.5b^2}, \quad r_2^{*2} = \frac{a^4 + 3a^2 b^2 + 3/8b^4}{a^2 + 0.5b^2}.$$

Therefore, the prescribed control law allows to change the angle θ into the required direction $(\theta \neq 0, \pi/2)$.

In Figs 4.3, 4.4 results of calculations of the change of parameters of motion of the system for the law of control $r = a - b \sin 2\varphi$ are presented for $a = 200$ km, $b = 0.1a$. The continuous lines show results of the numerical solution of the complete equations of motion. The high-frequency oscillations are presented only for the parameters p and L. For other parameters, because of their smallness and also because they are not essential, the high-frequency oscillations are smoothed. The dashed lines (Fig. 4.3) represent the solutions of equations (4.8). The results of the computations show good agreement of the solutions of the averaged and complete equations for the angle of nutation and the value of the moment of momentum. The essential deviations in the results of calculations of the angle of precession are explained by the significant change of velocity of the argument of the pericentre (Fig. 4.4).

For the law of control $r = a \pm b \sin 2(\nu - \psi)$ the equations similar to (4.8) look like

$$\frac{d\theta}{d\nu} = \mp \frac{3}{4} \sqrt{\frac{\mu}{p^3}} \frac{ab}{L} \sin \theta, \quad \frac{d\psi}{d\nu} = -\frac{3}{4} \sqrt{\frac{\mu}{p^3}} \frac{a^2 + 0.25b^2}{L} \cos \theta,$$

$$\frac{dL}{d\nu} = 0. \tag{4.9}$$

In Figs 4.5 and 4.6 (as in Figs 4.3 and 4.4) results of calculations of the change of parameters of the system for the law of control $r = a - b \sin 2(\nu - \psi)$ are presented for $a = 200$ km, $b = 0.1a$.

In this way the controllability of the orientation of the rotational relative motion of the system is shown. Moreover, the laws resulting in purposeful changes of the nutation angle can be based on tuning both the "internal" and "external" resonances.

Since the laws of control, considered above, were not constructed as optimal, apparently, the combination of control of "internal" and "external" resonances of the system will allow to increase the control efficiency and improve transient processes [127, 129].

4.3   Control of orbital motion

The possibility of control of the parameters of orbital motion of the system follows from fact that for an orientation of the system different from equilibrium the force of attraction has transversal and normal components (Fig. 4.7).

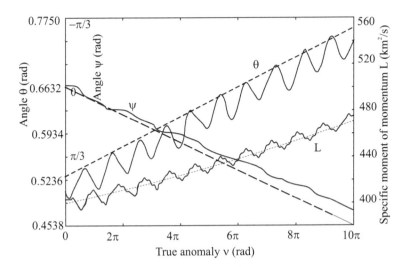

FIGURE 4.3
Variations of relative motion parameters at $r = a(1 - 0.1 \sin 2\varphi)$.

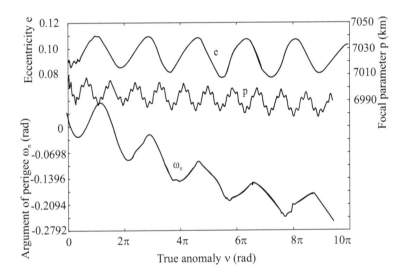

FIGURE 4.4
Variations of orbital motion parameters at $r = a(1 - 0.1 \sin 2\varphi)$.

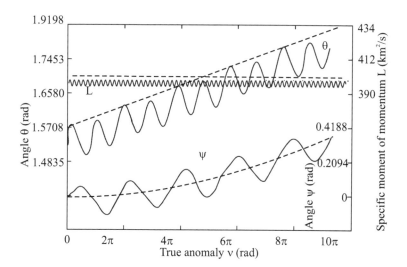

FIGURE 4.5
Variations of relative motion parameters at $r = a(1 - 0.1\sin(\nu - \varphi))$.

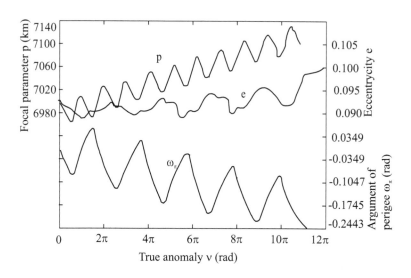

FIGURE 4.6
Variations of orbital motion parameters at $r = a(1 - 0.1\sin(\nu - \varphi))$.

Based on this obvious fact the various schemes of control of the orbital elements at the expense of internal forces can be constructed. These schemes can be based both on resonant tuning of frequencies of the change of length of the connection for systems with connections of variable lengths and on the use of an internal control moment for systems with rigid bars of constant length.

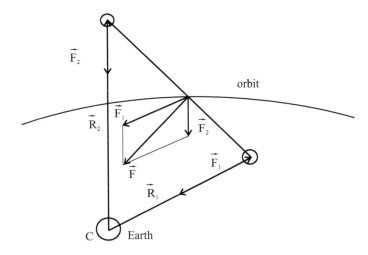

FIGURE 4.7
Geometrical parameters of a tether on an orbit.

The possibility of the change of orientation of the plane of the orbit follows from the possibility of change of the value and the orientation of the moment of momentum of relative motion. Since the ratio of velocity of the change of L to the angular orbital velocity is proportionally to $\sqrt{\mu/p^3 r^2}/L$, by virtue of (3.58) the essential change of the inclination of the orbit i is possible during $(p/r)^2/2\pi$ turns on the orbit. Velocities of control of the value of the angle Ω are similar as the moment of momentum of the motion of the mass centre moves on the surface of the cone with the half-opening angle equal to i:

$$\sin i = \frac{m_1 m_2}{M^2} \frac{L \sin \theta}{G'}.$$

The change of argument of the pericentre (angle between line of nodes and the line of earth centre to the perigee of orbit) and the eccentricity of the orbit is possible without change of the value and the orientation of the vector of moment of momentum of the orbital motion [20].

We consider possibilities of the change of the focal parameter for the motion of the system in the plane of orbit [89]. It is possible to write in this case equations (3.55), (2.87) in the form

$$\dot{L} = -\frac{3}{2} \frac{\mu}{R^3} r^2 \sin 2\lambda,$$

$$\dot{\lambda} = \frac{L}{r^2} - \dot{u},$$

$$\dot{p} = 3\frac{m_1 m_2}{M^2}\sqrt{\frac{p}{\mu}}\frac{\mu}{R^3}\sin 2\lambda,$$

$$\dot{u} = \frac{\sqrt{\mu p}}{R^2},$$

$$\dot{e} = \frac{3}{2}\frac{m_1 m_2}{M^2}\frac{\sqrt{\mu p}}{R^2}\left(\frac{r}{R}\right)^2\left[\sin\nu(1 - 3\cos^2\lambda) + \left(\cos\nu + \frac{e + \cos\nu}{1 + \cos\nu}\right)\sin 2\lambda\right],$$

$$\dot{\omega}_\pi = \frac{3}{2}\frac{1}{e}\frac{m_1 m_2}{M^2}\frac{\sqrt{\mu p}}{R^2}\left(\frac{r}{R}\right)^2\left[\cos\nu(3\cos^2\lambda - 1) + \left(1 + \frac{1}{1 + \cos\nu}\right)\sin\nu\sin 2\lambda\right]. \tag{4.10}$$

From equations (4.10) it is visible that the largest velocity of increase of the parameter p is reached at $m_1 = m_2$ and $\lambda = \pi/4$. Hence the dumb-bell with equal masses maintained in the state $\lambda = \pi/4$ at the expense of the control moment created by internal forces (for example, by a fly-wheel placed in the mass centre) is the most effective model for the increase of the focal parameter.

To average the equation (4.10) over u for $\lambda = \pi/4$, $m_1 = m_2$, as before, it is necessary to pass in the equations to the new independent variable u. Then we obtain

$$\frac{dL}{du} = -\frac{3}{2}\sqrt{\frac{\mu}{p^3}}r^2,$$

$$\frac{dp}{du} = \frac{3}{4}\frac{r^2}{p},$$

$$\frac{de}{du} = \frac{15}{16}\left(\frac{r}{p}\right)^2 e, \tag{4.11}$$

$$\frac{d\omega_\pi}{du} = \frac{3}{16}\left(\frac{r}{p}\right)^2.$$

The solution of equations (4.11) looks like

$$p = \left[p_0^2 + \frac{3}{2}r^2(u - u_0)\right]^{1/2}, \quad \omega_\pi = \omega_{\pi 0} + \frac{1}{4}\ln\left(\frac{p}{p_0}\right),$$

$$e = e_0\left(\frac{p}{p_0}\right)^{5/4}, \quad L = L_0 - 4(\sqrt{\mu p} - \sqrt{\mu p_0}), \tag{4.12}$$

where the value L characterizes the change of the moment of momentum of the fly-wheel.

The numerical solution of the complete equations of motion shows good agreement with the solution of the averaged equations (4.12). In Fig. 4.8, for the initial conditions $p = 6671$ km, $e_0 = 0.1$, $u_0 = 0$, $\omega_\pi = 0$ and for $a = 200$ km the change of the parameters of the orbital motion is shown. The solid lines show results of the numerical solution of the complete equations of motion, the dashed lines show results of the solutions (4.12).

Of course, the realisation of such a project is at the moment not feasible because it is not possible to create a rigid design in the extent in dozens of kilometers. The bending moment in the middle of a dumbbell $\alpha = \pi/4$ is equal to $3/2\mu r^2 m_1 m_2/(MR^3)$. For a five-kilometer dumb-bell with end masses of 100 kg this gives ≈ 1900 N m. The main technical obstacle to realise such a project consists in the impossibility to store inside the space vehicle the moment of momentum comparable with the value of the moment of momentum of the orbital motion of the vehicle. If it could be achieved, due to the moments of gravitational forces with a certain speed it would be possible to redistribute the moment of momentum between orbital and relative motion. The total moment of momentum of a system would not change at this. Realisation of such project would look fantastical externally: from a space vehicle the counterbalance is put forward and the vehicle begins to increase the velocity without any visible reasons. The picture is similar on the unreality to a fantastic witch flying in a mortar with a broom turned back.

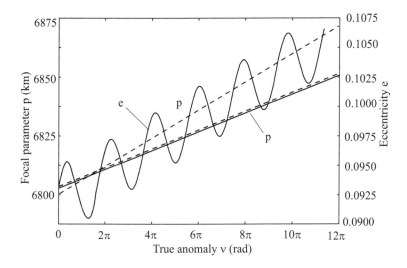

FIGURE 4.8
Variations of dumb-bell orbit parameters at $\lambda = \pi/4$.

The estimations from this model for the change of the parameters of the

TABLE 4.2

Change of the orbit elements
of the dumb-bell, $\lambda = \pi/4$.

r, km	$p - p_0$, km	e/e_0
0.1	0.00	1.000
1.0	0.07	1.001
10.0	7.06	1.011
100.0	672.50	1.101
200.0	2395.50	1.359

orbit for 100 turns around the attractive centre depending on length of the bar for $p_0 = 6671$ km are presented below.

From table 4.2 it is visible that the essential change of the parameters of the orbit can be reached for lengths of the bar of the rigid dumb-bell satellite starting from dozens of kilometers, a fact which essentially limits the possibilities of the practical realisation of this model of control.

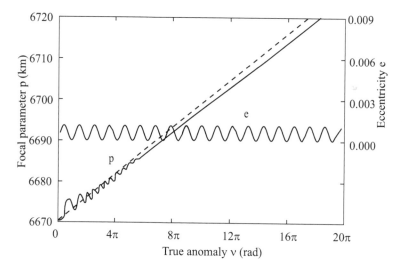

FIGURE 4.9

Variations of orbital parameters at $r = a(1 + 0.1 \sin 2\lambda)$.

The use of cables allows to considerably increase the extent of the system but complicates the transfer of the control moment. The idea of control of the parameters of the orbit of the TSS consists in the following: the distance between bodies is changed due to the action of internal forces. By this length change the angular velocity of rotation of the system is changed such that the system stays longer in the required orientation than in a state which results in an opposite effect for control. Hence, to increase p the change of length of the rotating system can be done according to the law $r = a + b \sin 2\lambda$. The solution of equations (4.10) averaged with respect to λ and u in this case differs from the solutions (4.12) that in the formulas for L, p, e r_1^{*2} is inserted instead of r^2, and for ω_π r_2^{*2} where r_1^{*2} and r_2^{*2} are the same expressions as in equations (4.8).

In Fig. 4.9 for the law of control $r = a(1 + 0.1 \sin 2\lambda)$, $a = 200$ km for the initial conditions $\lambda_0 = \pi/2$, $\dot{\lambda}_0 = -0.0001\,\mathrm{s}^{-1}$, $p_0 = 6671$ km, $e_0 = 0$, $u_0 = 0$, $\omega_{\pi 0} = 0$ the change of the focal parameter and eccentricity of the orbit obtained by numerical solution of the exact equations of motion (solid lines) and the solution of the averaged equations (dashed lines) is shown.

5

Deployment of Tethered Space Systems

One of the main modes of stabilisation of motion of a small TSS of length up to 1 km and total mass up to 100 kg is the stabilisation by rotation. In this connection a small TSS rotating about the mass centre is the most accessible version of a system for conducting the full-scale experiment. Small TSS stabilised by rotation can be also of independent interest. For example, if a rotating TSS may change its string length, the tension and angular velocity of the system can be used for studying the physics of the space plasma and the physics of the high atmosphere and magnetosphere. Rotating TSS can serve as the integrated sensor for studying of the effects of physical fields of the Earth. Rotating TSS with two tip bodies having various ballistic coefficients may be used for altitude measuring of atmospheric density. Here we consider the process of deployment from a spacecraft board of a small TSS stabilised by rotation. The physical parameters of the TSS have the following values: the distance l between bodies is of the order of 100 m, total mass m is about 20 kg. Angular velocity ω_c of the system is in a range $0.1 - 1\,\mathrm{s}^{-1}$. At an initial instant the plane of rotation of the system coincides with the orbital plane of the mass centre of TSS. It is supposed that TSS is fixed before deployment on the main spacecraft, which moves on a near-circular orbit practically coinciding with the demanded orbit of the mass centre of the TSS. Separation of bodies of TSS during its deployment is realised by means of spring thrusters, whose axes are located in the orbital plane at the separation instant.

The method of deployment of TSS, which is based on the use of gravitational and inertia forces is realised as follows. The first body separates along the axis OY in the direction opposite to the motion of the spacecraft with some initial relative velocity. In absence of any resistance to the extraction of the string, this body would transfer to an elliptical orbit with apogee in the separation point. If forces of resistance to extraction of the string are present, the picture changes a little. Nevertheless the body starts to go down and to come up to the spacecraft. In any instant it appears on the axis OX. At this instant, the second body separates along the axis OY also with the given relative velocity. As a result of deployment the system is rotating in the orbital plane.

In the deployment of a TSS with the directions of separation oriented under an angle, the bodies, in analogy with the previous method, are separated sequentially into the orbital plane. After separation of the first body with the given velocity, the second body separates also with the given velocity under the given angle to the direction of motion of the first body. During this process the trajectory of the first body does not deviate noticeably from a straight line in the orbital frame. The initial velocity of deployment must be chosen such that the kinetic energy of the first body has a sufficient value for extraction of the total length of the string. For the considered system the duration of this process is about some dozens of seconds in the presence of resistance to extraction. Different values of the forces of resistance have only an influence on the deployment time but not a qualitative one. In order to decrease the deployment velocity to zero in the final period of deployment a braking device is necessary.

Another realisation of the deployment is possible using an additional (third) body initially attached to the first body. If the additional body separates from the first one, the first body gets an incremental velocity directed under some angle to the X-axis. The second body separates from the main spacecraft with given velocity after the total deployment of the string. In such a way the system gets the prescribed angular velocity.

The simultaneous separation of three connected bodies by parallel thrusters with the subsequent division of the middle body into two parts by a spring is closely related to the previous method. There are two simultaneously rotating tethered systems here. An essential deficiency of both versions is the difficulty to guarantee that the velocities of the separated (or parted) bodies will be located in the orbital plane.

Rotation of the vector of initial velocity of the first body without separation of an additional mass is possible too. The first and second bodies of the system are fixed on two thrusters accordingly. The technological platform with the rotary lever established on it is fixed on the third thruster. After the first thruster operates and the first body gets a necessary velocity, it starts to move on around. After rotation of the lever on some angle, which should be chosen as much as possible close to right angle, the connection between the lever and the first body, which separates from the lever with the given velocity, is broken. Then the string is unwinding to total length. After this, the second body separates. This process may also be realised, for example using instead of the lever a flexible string. Essentially, the scheme will be the same.

5.1   Deployment with prescribed final motion

Here the approach to develop control algorithms based on the assignment of certain limitations for the trajectory of deployment in the phase space is described. It also allows to design the algorithms adapted to specific requirements of the final motion.

5.1.1 Prescribed constraints on phase variables

Constraints on velocities of separation, on angular direction of separation, and also on the possibility of obtaining the final rotation velocity of the tether in the scheme of TSS deployment described above are natural, because engineering realisation of all methods of deployment is related to the use of mechanical thrusters.

Equations of motion of a tether system with a straight-line cable are well-known [20]. In dimensionless variables x, y (x is coordinate along a local vertical) in a plane of circular orbit these equations look like

$$\ddot{x} - 2\dot{y} - 3x = -f\frac{x}{l}, \quad \ddot{y} + 2\dot{x} = -f\frac{y}{l}, \quad l = \sqrt{x^2 + y^2}, \tag{5.1}$$

f is the dimensionless force of tension of the cable

$$f = \frac{P}{m\omega_0^2 l}. \tag{5.2}$$

Selection of the law of change of f provides the possibility of control of motions of the system (5.1), including the process of deployment. The variety of types of motion, which are turning out by various selections of the function f raises the question on the description of all sets of such motions and possible limitations on them. Such limitations can be obtained, eliminating f from equations (5.1). Having performed such eliminating, we find the condition

$$x\ddot{y} - y\ddot{x} + 2x\dot{x} + 2y\dot{y} + 3xy = 0. \tag{5.3}$$

on motions of the system (5.1) for any control law for the tension force (together with length) of the cable. It is easy to prove the proposition:
Any motion $x(t), y(t)$, satisfying relation (5.3) is for $x^2 + y^2 \neq 0$ the solution of the system (5.1) for suitable selection of the function f.

In fact, let on some segment of motion $t_0 \leq t \leq t_1$ the coordinate $x \neq 0$. Then for known $x(t), y(t)$ from the first equation (5.1) we find f. Thus by virtue of (5.3) the second equation (5.1) is applied also. The proof in the case $y \neq 0$ is similarly carried out.

Certainly, so that the motion found actually could be realised, it is necessary that the appropriate tension of the cable was positive.

The specified property enables to find a set of various motions of the system among which some useful for deployment may also appear. One of possible ways of construction of such motions consists in the following. We impose at the system phase variable some limitation of the kind

$$F(x, y, \dot{x}, \dot{y}) = 0. \tag{5.4}$$

Differentiation of this relation with respect to time gives

$$\frac{\partial F}{\partial \dot{x}}\ddot{x} + \frac{\partial F}{\partial \dot{y}}\ddot{y} + \frac{\partial F}{\partial x}\dot{x} + \frac{\partial F}{\partial y}\dot{y} = 0. \tag{5.5}$$

Equations (5.3) and (5.5) under condition of

$$\frac{\partial F}{\partial \dot{x}}x + \frac{\partial F}{\partial \dot{y}}y \neq 0 \qquad (5.6)$$

uniquely allow to express $\ddot{x}, \ddot{y}$ through state variables and, thus, determine a system of ordinary differential equations of the second order for x, y. This system should satisfy relation (5.4). Then its solutions give the family of motions of the tether system lying on variety (5.4).

5.1.2   Prescribed trajectory

Special interest represents the case if the constraint (5.4) looks like

$$F(x, y) = 0, \qquad (5.7)$$

that corresponds to the task of a trajectory in the plane (x, y). The condition (5.6), and the additional linear relation between $\ddot{x}$ in this case is not satisfied and $\ddot{y}$ turns out by double differentiation on time of (5.7). For unique solvability of the obtained system of equations with respect to $\ddot{x}$ and $\ddot{y}$ fulfilment of the condition

$$\frac{\partial F}{\partial x}x + \frac{\partial F}{\partial y}y \neq 0. \qquad (5.8)$$

is necessary. This condition is fulfilled identically only on straight lines

$$ax + by = 0, \qquad (5.9)$$

but it is known that if $a^2 + b^2 \neq 0$, such straight-line trajectories may turn out at exponential deployment in cases when the tension force is proportional to the length also [25, 56]. Thus, control of a tethered satellite with the help of a cable allows to construct any trajectories in the plane of orbit, except straight lines $x = 0$ and $y = 0$ and separate points on other trajectories. Certainly, only those of them to which corresponds a positive tension of the cable are actually realised.

5.1.3   Monotonous tether feed out

Let us consider one example, which is of interest at the final stage of the process of deployment. Not all known algorithms give quite satisfactory results right at the end of deployment. In some of them there is offered to terminate deployment by a jerk (sharp braking), in other algorithms after stop of deployment there are significant pendulous oscillations of the system, and even in the most successful (for example, [99]) for full extinction of pendulous oscillations at the end of deployment it is necessary to use retrieving motion of cable. In this connection arises the question: what control should be at the final stage of deployment such that the subsatellite reaches a terminal point with zero velocity and that the approach to this point occurs monotonously, without change of sign of the velocity?

Let the terminal point have dimensionless coordinates $(1,0)$. Let us consider a possible motion on the straight line

$$y = k(x-1), \tag{5.10}$$

which passes through this point. Having substituted this relation in (5.3), we find an equation for x:

$$k\ddot{x} + 2(1+k^2)x\dot{x} - 2k^2\dot{x} + 3kx(x-1) = 0. \tag{5.11}$$

Leaving out the question the general integration of this equation, we try to find an interesting partial solution. We determine a function G as follows

$$G = k\dot{x} + (k^2+1)x^2 + ax \tag{5.12}$$

and select constant k, a, s such that from (5.11) it follows

$$\dot{G} = sG \tag{5.13}$$

From this condition we find:

$$a = -\frac{\sqrt{13}-1}{2} \approx -1.30278, \quad k = \pm\sqrt{\frac{\sqrt{13}-3}{2}} \approx \pm 0.55025,$$

$$s = \frac{3k}{a} \approx \mp 1.26710. \tag{5.14}$$

$$G = G_0 e^{st}. \tag{5.15}$$

Substituting this value of G in (5.12), we find for x a nonautonomous equation of the first order. The simplest case which we also use turns out at $G_0 = 0$. The equation (5.12) in view of equality (5.14) becomes

$$k\dot{x} + (k^2+1)x(x-1) = 0 \tag{5.16}$$

and is easily integrated. The solution looks like

$$x = \frac{e^{\varphi}}{2\cosh\varphi}, \quad y = -\frac{ke^{-\varphi}}{2\cosh\varphi}, \quad \varphi = \frac{k^2+1}{2k}(t+t_0), \tag{5.17}$$

where t_0 is an undefined constant. From formulas (5.17) follows that the positive value k corresponds to deployment: at $t \to +\infty$ $x \to 1$. It is easy to check that for motion (5.17) under conditions $k > 0$ and $0 < x < 1$ the tension of the cable is positive and $f \to 3$ for $x \to 1$. The obtained result shows the possibility of monotonous approach of the final state in the deployment of a tether of the system.

5.2   Deployment of a rotating TSS

One operational mode of motion of a TSS is such that the mass centre of the system moves on the given orbit, the string connecting the body is under tension and the system rotates in the orbital coordinate system in the orbital plane. Thus it is supposed: 1) the system is fixed before deployment on the main vehicle, which moves on orbit practically coinciding with the required orbit of the mass centre of the system; 2) separation of bodies of the system during its deployment is performed with the help of spring pushers, the axes of the springs are located in the plane of orbit at the beginning of separation of the bodies.

To concretise the problem, we assume that the parameters of the rotating tether system have the following values: distance l between bodies about 100 m, mass m of each of the bodies is within the limits of $1-10$ kg, angular velocity ω_c of the rotation of the system is equal $0, 1-1\,\mathrm{s}^{-1}$. The orbit of the main vehicle is supposed to be circular.

5.2.1   Deployment due to gravitational and inertial forces

Deployment of the tethered system happens by the realisation of motion of the first body directionally, close to a trajectory of the free motion, i.e., only under the effect of gravitational forces. The scheme of deployment of the system is presented in Fig. 5.1. We connect to the spacecraft a frame of reference $Oxyz$ with the origin in the mass centre of the spacecraft. The axis Ox is directed along the position vector of the orbit of the spacecraft, the axis Oy is directed along the tangent to the motion of the spacecraft in the orbital plane, and the axis Oz supplements the frame of reference to right-handed one.

The first body separates along the y-axis in the direction opposite to the spacecraft motion with the relative velocity $\vec{V}_{depl}^{init}$ and moves along the shaped line. After deploying this body along the y-axis, the second body separates along the x-axis with the relative velocity $\vec{V}_{rot}$. As result of deployment the system is rotating in the orbital plane. We consider that separated bodies are the mass points coinciding in separation instant with the mass centre of the spacecraft. According to [20], the current coordinates x, y of the separated body for a circular orbit of the spacecraft in case of a plane motion satisfy the equations:

$$x'' - 2y' - 3x = 0, y'' + 2x' = 0. \tag{5.18}$$

Primes in (5.18) mean derivatives on the dimensionless time, $\tau = \omega_o\, t$, ω_o is an absolute value of angular velocity of motion of the mass centre of the spacecraft on orbit. Having integrated these equations, we obtain [20]

$$x = 2c_1 + c_2 \sin\tau + c_3 \cos\tau, y = c_4 - 3c_1\tau + 2c_2 \cos\tau - 2c_3 \sin\tau, \tag{5.19}$$

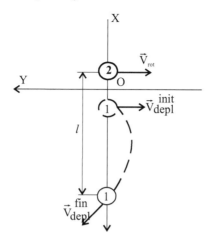

FIGURE 5.1
Deployment of a system with the use of central force.

where arbitrary constants c_1, c_2, c_3, c_4 may be expressed in terms of initial (at $\tau = 0$) values of coordinates and their first derivatives (x_0, y_0, x_0', y_0').

Having determined the integration constants for $x_0 = 0, y_0 = 0, x_0' = 0, y_0' \neq 0$ we obtain

$$x = y_0' \cdot 2 \left(\cos \tau - 1\right), y = y_0' \cdot \left(3\tau - 4\sin \tau\right), \qquad (5.20)$$

where y_0' is the value of the relative velocity of separation of the body in the dimensionless time τ, related with V_0 by the ratio

$$y_0' = \omega_o^{-1} \cdot V_0. \qquad (5.21)$$

V_0 is the relative velocity of separation of the body in time t.

From the condition $y\left(\tau^*\right) = 0$ the dimensionless time of deploying the first body in free motion on the X-axis (deployment time) is determined: $\tau^* = 1.276$. This constant conforms to transformation of zero initial conditions of variable y to initial conditions for the dimensionless time.

From (5.19), (5.20) for $x\left(\tau^*\right) = l$ we obtain the expression of velocity of separation of the first body $V_{\text{depl}}^{\text{init}}$

$$V_{\text{depl}}^{\text{init}} = 0,704\omega_{\text{orb}}l.$$

For orbits of an altitude of about $500\,\text{km}$, at $\omega_{\text{orb}} \approx 1, 2 \cdot 10^{-3}\text{s}^{-1}$ and $l \approx 100\,\text{m}$ we have

$$V_{\text{depl}}^{\text{init}} \approx 0,08 \text{ m/s} \qquad (5.22)$$

Dimensionless time of deployment is always equal 1,276. Absolute time t_{depl} is equal:

$$t_{\text{depl}} = \omega_{\text{orb}}^{-1} \cdot \tau^* \approx 17,7min.$$

The angular velocity of the relative rotation of the system is determined by the formula

$$\omega_c = \frac{V_{rot} + \dot{y}\,(\tau^*)}{l}, \tag{5.23}$$

where V_{rot} is the velocity of separation of the second body; $\dot{y}\,(\tau^*)$ is the transversal component of the velocity of the first body relative SV (Fig. 5.1) in absolute time to separation moment of the second body.

When we find $\dot{y}\,(\tau^*)$ ω_{orb} with the help of the second equality (5.19), we obtain for $\omega_{orb} = 1,2 \cdot 10^{-3}\mathrm{s}^{-1}$:

$$\omega_c \approx \frac{V_{rot}}{l} + 1,5 \cdot 10^{-3}. \tag{5.24}$$

The last ratio shows that for the required values ω_c with value about 0.1 s^{-1} the second item has no practical meaning. Hence, at first sight, the attractive capability to increase the angular velocity of rotation of the tether system due to the use of gravitational forces in the given situation does not give a practical advantage. At the same time, determined by equation (5.22) velocity $V_{\mathrm{depl}}^{\mathrm{init}}$ as show results of experiments with nominal spring pushers, is realised with big errors of the given masses of bodies that results in oscillations of the distance between bodies at the moment of deploying of the first body on straight vertical.

We note that the case is considered where the first body has negligible small effects from the string. We estimate the order of force of resistance against unwinding of the string, appreciably influencing the trajectory of the first body. The force F_{res}, completely extinguishes the kinetic energy of the relative motion of the body of mass m at length l with the initial velocity $V_{\mathrm{depl}}^{\mathrm{init}}$. This may be found from equating kinetic energy and the work of the force, given by $m\left(V_{\mathrm{depl}}^{\mathrm{init}}\right)^2/2 = F_{res} \cdot l$.

For $l \approx 100\,\mathrm{m}$ and $\omega_{orb} \approx 1,2 \cdot 10^{-3}\mathrm{s}^{-1}$, we obtain

$$\frac{F_{res}}{m} \approx 0,36 \cdot 10^{-4}\frac{\mathrm{m}}{\mathrm{s}^2}.$$

Numerical integration of equations of motion of the first body taking into account F_{res} shows that for $F_{res}/m = 4 \cdot 10^{-5}\mathrm{m\,s}^{-2}$ the distance of the first body from the second one decreases $\sim 15\%$ in comparison with $F_{res} = 0$ at the instance of deployment along the local vertical, and for $F_{res}/m = 8 \cdot 10^{-5}\,\mathrm{m\,s}^{-2}$ — on $\sim 30\%$.

Thus, the carried out analysis shows that deployment of the tethered system in the considered way by the organisation of motion of the first body on a trajectory, close to the trajectory of free motion, assumes the decrease of the force of resistance of unwinding of the string of below 0.001 of gram for a kilogram of mass of the first body. This requirement should be executed to provide parameters of motion of a tether system close to computational values. Otherwise, wide scatter of ω_c takes place at unstable l that follows from (5.24). Technical difficulties of guaranteeing so small values of the resistance to unwinding of a string are obvious. Hence this essentially reduces the practical expediency of application of the considered way for deployment of rotating systems with the given parameters.

5.2.2 Deployment along an inclined direction to the local vertical

The scheme is shown in Fig. 5.2. Bodies, as in the previous way, are separated consistently in the plane of orbit. After separation of the first body with velocity V_{depl}, the second body is separating with the velocity directed under an angle ξ to the velocity V_{rot}. It is supposed, as in the first way, that the bodies are connected by a string of required length, which in its initial position is packed and begins to unwind from the moment of separation of the first body.

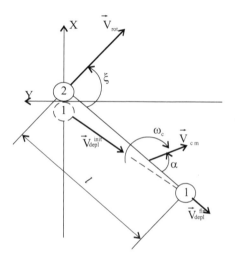

FIGURE 5.2
Deployment of a system with inclined axes of devices of separation against the local vertical.

For the analysis we estimate first of all time t_p, which is the duration of pushing apart of the two bodies. From formula (5.19) it is visible that the trajectory of the first body can be considered as straight-line for $\cos \tau \approx 1, \sin \tau \approx \tau$. These approximations are applied with accuracy of one tenth of a percent, if they do not exceed 0.01. For $\omega_{\text{orb}} = 10^{-3}\text{s}^{-1}$ the value t_p should be chosen, in connection with what was said, about several tens of seconds.

Let us write the formula for the angular velocity of rotation of the tether system ω_c. The moment of momentum of the system at rotation with respect to its own centre of mass after separation of the second body is determined by the expression

$$\vec{G} = \frac{m_1 \cdot m_2}{M} \vec{r} \times \left(\vec{V}_{\text{depl}} + \vec{V}_{rot} \right), \tag{5.25}$$

where m_1, m_2 are the masses of the first and second body accordingly; $M = m_1 + m_2$; $\vec{r}$ is the radius-vector, which has been drawn from the centre of mass of the second body to the centre of mass of the first one.

Neglecting external forces (gradient of gravitational forces, aerodynamics etc.), it is possible to set for an initial stage of motion after deployment

$$\vec{G} = const. \tag{5.26}$$

In operational mode of motion of the bodies of the tether system may be considered as point masses that allows to write for the absolute value of the moment of momentum

$$\left| \vec{G} \right| = \frac{m_1 \cdot m_2}{M} l^2 \cdot \omega_c, \tag{5.27}$$

where l is length of the system.

On the basis (5.23), (5.27) and taking into account that $l = \left| \vec{V}_{\text{depl}} \right| \cdot t_p$ we may write

$$\omega_c = \frac{\left| \vec{V}_{\text{depl}} \right| \cdot \left| \vec{V}_{rot} \right| \cdot t_p}{l^2} sin\xi. \tag{5.28}$$

The obtained formula shows that non-simultaneous separations of bodies are essentially necessary in the considered way since at $t_p = 0$ we obtain $\omega_c = 0$. In this case the line connecting the mass centres of the bodies moves in parallel, not changing the angular position (straight-line trajectories of each of the bodies are considered).

At the separation instant of the second body the first one moves away from the main vehicle with the distance

$$l_0 = \left| \vec{V}_{depl} \right| \cdot t_p.$$

Then $\vec{V}_{depl}^{init} = l_0/t_p$.

Substituting this in (5.28), we obtain

$$\omega_c = \frac{\left| \vec{V}_{rot} \right| \cdot l_0}{l^2} \sin \xi. \tag{5.29}$$

This relation may be used for selection of parameters of the TSS system for deployment. Therefore, for achievement of maximum angular velocity and the fullest use of energy of the devices of separation it is necessary to separate the second body in a direction perpendicular to the direction of separation of the first body ($\xi = \pi/2\,\text{rad}$) and to provide the greatest possible distance l_0 of the first body at the moment of separation of the second one.

The vector of initial velocity of the mass centre of the system $\vec{V}_{cm}$ may be found from the formula

$$\vec{V}_{cm} = \frac{m_1 \cdot \vec{V}_{depl}^{fin} + m_2 \cdot \vec{V}_{rot}}{M}, \tag{5.30}$$

where $\vec{V}_{depl}^{fin}$ is the velocity vector of the first body at the moment of separation of the second one (Fig. 5.3).

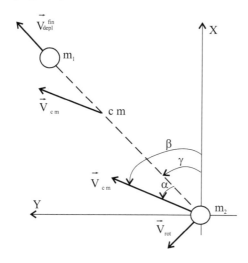

FIGURE 5.3
Definition of the velocity of the mass centre of the system.

Considering $V_{depl}^{fin} = 0$ which may be achieved with the braking device, we obtain for the value and the inclination of the velocity vector $\vec{V}_{cm}$ of the mass centre of the system from equation (5.30):

$$\left|\vec{V}_{cm}\right| = \frac{\varepsilon}{\varepsilon + 1}\left|\vec{V}_{rot}\right|,$$
$$\beta = \gamma + 1,$$

where $\varepsilon = m_2/m_1$; β is the angle between $\vec{V}_{cm}$ and the x axis. By selection β is possible to provide a trajectory of the mass centre of the system close to the required one.

Let us add quantitative estimations. From (5.29) at $\xi = \pi/2\,\mathrm{rad}$ and $l_0 = l$ we obtain:

$$\omega_c = \frac{\left|\vec{V}_{rot}\right|}{l}.$$

At $V_{rot} = 12\,\mathrm{m\,s^{-1}}$ (the maximum velocity provided with the nominal pusher) we have

$$l = 100\,m, \quad \omega_c = 0,12\,\mathrm{s^{-1}},$$
$$l = 20\,m, \quad \omega_c = 0,6\,\mathrm{s^{-1}}.$$

Let us turn to the estimation of the effect of resistance force to unwinding. The value of velocity $V_{\text{depl}}^{\text{init}}$ should be chosen such that the kinetic energy of the first body is sufficiently large for full deployment of the string in the presence of resistance against extraction. To find the precise value of this velocity (a minimum value must be exceeded) is not so critical, since damping of the residual velocity is supposed to occur at the end of string deployment. In detail the question is considered in [20].

Let us consider result of power estimations. For a resistance to unwinding of $10\,g$ for a kilogram of mass of the first body at $l = 100\,\text{m}$ the minimal initial velocity of the first body must be equal $4.5\,\text{m\,s}^{-1}$ that is reached by the nominal pusher.

Given estimations confirm feasibility of this consideration as the basis for the technical realisation of deployment of a system of TSS.

5.2.3 Deployment with changing of velocity of bodies after separation

The analysis of the way of deployment carried out above with rotation of the axes of the devices of separation shows that the rotation of the axes is the essential factor. Two effects are basic here: the full unwinding of the string until the moment of separation of the second body and separation of the second body in the direction, perpendicular to the deployed system (the task is realisation of a preset value ω_c). Thus the initial orientation of the velocity of the first body and the further changes of spacecraft attitude are not so important. We consider the case when the axes of both pushers are parallel, and the direction of velocity of the first body in addition changes after shooting. We consider two versions.

Let us address Fig. 5.4. The first body with mass m_1, is separated along the y axis with velocity $\vec{V}_1^{\text{orig}}$. A bit later after the first body with velocity $\vec{V}_m$ the auxiliary mass m is separated (for example, by the spring pusher). As a result the first body obtains the increment of velocity $\vec{V}_1^{\text{add}}$ which, added to $\vec{V}_1^{\text{orig}}$, provides to this body the velocity $\vec{V}_1^{\text{init}}$ under an angle α to the x axis. After deployment of the total string the second body is separated from the main vehicle with the velocity $\vec{V}_{rot}$, therefore the system gets the angular velocity

$$\omega_c = \frac{\left|\vec{V}_{rot}\right|}{l} \cdot \cos\alpha.$$

Numerical estimations show that the tether system with certain parameters determined by the input data (ID) may be created by the simultaneous separation of three connected bodies by parallel pushers with the subsequent separation of the middle body with the help of a spring. Two rotating systems are created here simultaneously.

The essential problem of both versions is the difficulty of maintenance of

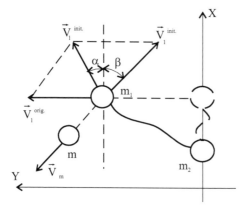

FIGURE 5.4
Deployment of a tether with additional mass separation.

the arrangement of velocities of the separated (or divided) masses in the plane of orbit.

The version of turn of a vector of initial velocity of the first body without shooting of an additional mass (Fig. 5.5) is interesting. On pushers *1T* and *2T* the first and second bodies of a system accordingly are fixed. On the pusher *3T* the technological platform is fixed with the rotary lever established on it.

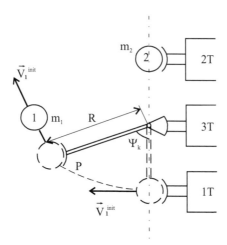

FIGURE 5.5
Separation of the first body with use of a rotating lever.

After actuation of the pusher *1T* and giving to the first body the velocity $\vec{V}_1^{\text{orig}}$, it begins to move on a circle of radius r with centre in point O. After turn of the lever angle ψ_k, which should be chosen so close to $\pi/2$ as far as design reasons allow, the connection between the lever and the first body, which is separated from the lever with velocity $\vec{V}_1^{init}$, is broken off. Then the unwinding of the string, then the second body is separated.

Another version assumes the use of the flexible connection instead of the lever, for example, a string. Essentially the scheme is the same, but some new design features appear.

This version is free from the above mentioned problem connected to the difficulties of maintaining the turn of velocity $\vec{V}_1$ in the orbit plane since the axis of turn of the lever is rigidly fixed relative to the SV.

Let us focus on the version of change of angular velocity of the system after its full deployment by the change of distance between the bodies. The maximum velocity of the body provided with the nominal pusher is equal $12\,\mathrm{m\,s^{-1}}$, whence according to (5.30), for $l = 100\,\mathrm{m}$ we have $w_c = 0,12\,\mathrm{s^{-1}}$. This is the maximum value, which may be obtained in these conditions without additional actions.

From (5.26) and (5.27) it is possible to obtain a relation for the freely moving system after full separation from the main vehicle

$$\frac{\omega_1}{\omega_2} = \left(\frac{l_2}{l_1}\right)^2, \tag{5.31}$$

where ω_i, l_i are angular velocities and lengths of the system accordingly.

Let l_1 be the initial length of the deployed system and l_2 its nominal length. The system deployed up to length l_1 has angular velocity

$$\omega_1 = \frac{V_{max}}{l_1}. \tag{5.32}$$

Thus the maximum achievable angular velocity without additional retraction of bodies of the system is equal

$$\omega_{pas}^{max} = \frac{V_{max}}{l_2}. \tag{5.33}$$

Reducing the length of the tether after deployment to length l_2, we obtain angular velocity

$$\omega_{act}^{max} = \omega_1 \left(\frac{l_1}{l_2}\right) = \omega_{pas}^{max} \left(\frac{l_1}{l_2}\right).$$

Thus, by deploying a system up to length l_1, and then by reducing distance between bodies up to l_2, it is possible to increase the maximum angular velocity l_1/l_2 times in comparison with the deployment to the necessary length at once. This is convenient, since the retrieval of bodies does not change the direction of ω_c.

The carried out analysis of the different ways of deployment of the system offers for design a version with turning of the axes of pushers by $\pi/2$ and consecutive separation of the bodies.

5.3 Deployment of three elastically tethered bodies in the centrifugal force field

5.3.1 Physical model

One of possible implementations of small space tethered systems (about several dozens of meters) represents a system of three bodies connected by an elastic tether. The masses of the end bodies of such a system practically determine the total mass of the system (see Fig. 5.6), the central body is used as an auxiliary one during deployment. The tether and the braking devices are located in this body yielding the required mode of deployment.

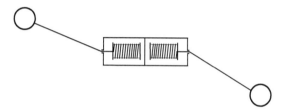

FIGURE 5.6
Physical model of the tethered three bodies.

The simplest and cheapest implementation, concerning the mathematical description of deployment of such a system in orbit is deployment under the action of the centrifugal forces resulting from their initial rotation in the plane, passing through the mass centres of the end bodies. For this purpose two pushers (for example, springs) should be mounted on the space vehicle, which by command from the Earth or from onboard the space vehicle push the departing system into a rotational motion in the given plane.

The initial angular velocity should be sufficiently large to ensure on one hand deployment of the system under the action of the centrifugal forces, and on the other hand to reach the prescribed angular velocity of the tether, prescribed by the operation conditions, after its full deployment.

The mathematical description of the process of pushing away of the system as a rigid body from the space vehicle is simple enough such that we do not enter into it. Next the connections, which are used to make the three body system as one rigid body before deployment, are removed. This is the instant of separation of the bodies due to the pushers which become effective now. The system is now a free system of three bodies connected by two tethers, performing a relative motion under the action of centrifugal and gravitational forces.

To construct the mathematical model of the system, we consider the bodies as rigid. We assume that the deploying strings at the points of exit from the middle body can be braked with the help of Coulomb and viscous friction forces. Since the gravitational forces and the centrifugal forces of orbital motion are much smaller than the centrifugal forces of attitude motion, we can neglect the moments of both of them due to the small sizes of the system because these moments cannot noticeably change the moment of momentum obtained by the system at repulsion.

Because the size of the tether under consideration is essentially less than the orbit size, the problem may be considered in the "limited" statement when the equations of motion of the mass centre of the tether are independent of the attitude motion. Further, we concentrate the attention on the study of the attitude motion of the system.

The mass centre of the system moves in space relative to the pushing device that yields the initial values of its state variables. The motion of the deployed system occurs inside some area of space, in which there should not be any restriction to its motion. The approach to construction of boundaries of such an area is similar to the one described in [126].

The strings are extracted from the storage containers located in the middle body under the action of centrifugal forces. Simultaneously some friction forces act on the strings due to the braking devices.

The friction forces should be adjusted such that when the end bodies reach their design positions with respect to the central body there should not occur an elastic impact, large enough to break the cable. However, some elastic oscillations along the cable can arise which will be damped under the action of forces of structural damping.

5.3.2   Mechanical model

The considered assumptions about the character of the process of the tether deployment allow to construct the adequate mechanical model shown in Fig. 5.7.

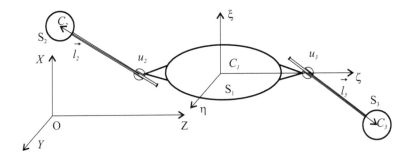

FIGURE 5.7
Mechanical model of the tethered three bodies.

Here $OXYZ$ is the absolute frame of reference, S_1 is the central body with the frame of reference $C_1\xi\eta\zeta$ fixed in it, S_2, S_3 are the end bodies, u_2, u_3 are the centres of the generalized hinges. Each of these hinges generally provides four relative degrees of freedom for the proper end body: three angular and one linear, which corresponds to moving out the elastic rods, simulating the strings. Such free system of bodies in case of the inextensible string can be described within the limits of the classical multibody theory [126].

Under consideration of elastic deformations of the string we take advantage of some simplifying assumptions. First of all, we consider that the string is absolutely flexible. This permits to consider the points of exit of the strings from the central body as spherical hinges, resulting in simply supported boundary conditions. The axial stretching deformation of the string is described by the linear Hook's law. We neglect the mass of the strings in comparison with the masses of the bodies of the system. We assume that the friction forces in the points of exit of the strings from the central body are adjustable either as a function of time, or as a function of nominal length of the string, which has been pulled out from the container. We assume also that pulling out the string from the container happens monotonically, without stops, while all string is pulled out. This assumption, quite justified from the practical point of view, allows essentially to simplify taking into account the elastic deformations of the string at deployment, as in this case its tension is determined only by the friction force in the braking device, and its stiffness on stretching is determined by the length of the pulled out part of the string.

In the tether deployment one can select three stages of motion:

1. A stage of pulling out of the strings from the containers. The force of tension of each string is determined by force of resistance to pulling out of the string at this stage.

2. A stage of motion of the deployed tether with the stretched strings. In this case the force of tension of each string is determined by its relative elongation.

3. A stage of motion of the deployed tether with unstrained strings (or a string). In this case the corresponding forces applied to the bodies of the system are equal to zero.

5.3.3 Mathematical model

For the description of the dynamics of such a system it is possible to use many approaches. From the point of view of simplicity of the mathematical description and of writing the computing program for the numerical simulation of deployment, the use of Newton's second law and the theorem of change of moment of momentum is most convenient. The system shown in Fig. 5.8 can be presented as three free bodies with the applied external forces determined by the internal tension forces of the strings. The state of this system can be uniquely determined by the following variables: X_1, Y_1, Z_1 are the coordinates of the mass centre of the central body in an absolute frame of reference; $\varphi_1, \varphi_2, \varphi_3$ are the Bryant's angles of orientation of the central body in the absolute frame of reference; $X_2, Y_2, Z_2, X_3, Y_3, Z_3$ are the absolute coordinates of the point masses S_2 and S_3 accordingly. Hence, the system has twelve degrees of freedom.

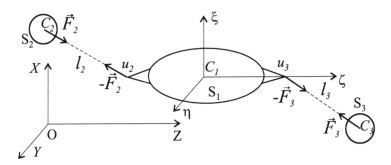

FIGURE 5.8
System of three free bodies.

To define the direction of the forces applied to the bodies, one can use auxiliary variables $X_2^u, Y_2^u, Z_2^u, X_3^u, Y_3^u, Z_3^u$. These are the absolute coordinates of the points u_2 and u_3 of exits of the strings from the container. They can be expressed in terms of the basic variables $X_1, Y_1, Z_1, \varphi_1, \varphi_2, \varphi_3$. Direction cosines of the unit vectors $\vec{l}_{k0}$ $(k = 2, 3)$ in direction of forces (see Fig. 5.8) may be found as

$$\vec{l}_{k0}(t) = \left\{ \frac{X_k - X_k^u}{l_k(t)}, \frac{Y_k - Y_k^u}{l_k(t)}, \frac{Z_k - Z_k^u}{l_k(t)} \right\}.$$

Here $l_k(t)$ is the current length of the corresponding string. If the distance is less than the deployed length of the string, then the string is not stretched and the corresponding forces are equal to zero.

The equations of motion of the tether at all stages of its deployment can be written as follows:

$$m_1 \ddot{\vec{R}}_1 = -\vec{F}_2 - \vec{F}_3, \quad m_2 \ddot{\vec{R}}_2 = \vec{F}_2, \quad m_3 \ddot{\vec{R}}_3 = \vec{F}_3, \tag{5.34}$$

$$\Theta^{C_1} \dot{\vec{\omega}} + \vec{\omega} \times \Theta^{C_1} \cdot \vec{\omega} = \vec{m}^{C_1}. \tag{5.35}$$

Here m_k and $\vec{R}_k$ are masses of the bodies S_k and their position vectors in the absolute frame of reference respectively, $\vec{F}_2, \vec{F}_3$ are the tension forces of the strings directed from the end bodies towards the central body, Θ^{C_1} is the tensor of inertia of the central body with respect to its mass centre, C_1, $\vec{\omega}$ is the vector of angular velocity of the central body, $\vec{m}^{C_1}$ is the vector of the moment acting on the central body due to the tension forces of the strings.

If one supplements the system (5.34), (5.35) with the kinematic equations and initial conditions, the initial value problem may be formulated.

The elastic forces created in the strings due to the motion of the end bodies can be determined as follows. Until the instant of termination of deployment $t = T_k$, the string tension is completely determined by the braking force at the point of exit of the string from the container:

$$\vec{F}_k = \vec{F}_k^{br}. \tag{5.36}$$

The tension force at $t > T_k$ when the string is stretched $(l_k(t) > L)$ can be determined as follows:

$$\vec{F}_k = -\vec{l}_{0k} \frac{l_k(t) - L}{L} EF \quad \forall l_k(t) > L. \tag{5.37}$$

In the absence of stretching, $\vec{F}_k \equiv 0$.

To formulate the initial conditions of the problem we consider that the initial rotation happens as the result of pushing the tethered bodies from the space vehicle with the help of two parallel forces resulting in different impulses. In particular, the impulse of one of the forces can be equal to zero. Besides, we assume that the instant of eliminating the initial connections coincides with the instant of termination of the action of the pushing forces. Thus the initial angular velocity has the value $\vec{\omega} = \vec{\omega}_0$, and relative velocity of the mass centre has the value $\vec{v} = \vec{v}_0$. The mass centre of the tether is located in the point C_0, which we fix in the frame of reference connected to the space vehicle. We connect the coordinate frame $C_0 x_0 y_0 z_0$ with this point such that the axis $C_0 x_0$ is directed in the direction of motion of the mass centre of the tether, the axis $C_0 z_0$ is collinear to the vector $\vec{\omega}_0$ and this frame of reference is non-rotating in the absolute frame of reference. This assumption is justified because the duration of deployment is substantially smaller than the orbital period.

The frame of reference $C_0 x_0 y_0 z_0$ is inertial (absolute) also because it moves with respect to the inertial frame of reference with constant velocity. Without loss of generality in posing the problem, we can consider the initial state of the system in the frame of reference $C_0 x_0 y_0 z_0$ as it is shown in Fig. 5.9.

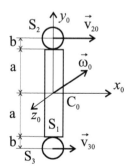

FIGURE 5.9
Initial state of the system before deployment.

For studying only the attitude motion, one can introduce the nonrotating frame of reference $Cxyz$, connected with the mass centre of the tether. Its axes are parallel with the proper axes of the frame of reference $C_0x_0y_0z_0$. Let $\vec{r}_k = \{x_k, y_k, z_k\}$ $(k = 1, 2, 3)$ are position vectors of the points C_1, C_2, C_3 in this frame of reference. Then the Cauchy problem can be written as follows.

The equations of motion:

$$m_1\ddot{\vec{r}}_1 = -\vec{F}_2 - \vec{F}_3, \quad m_2\ddot{\vec{r}}_2 = \vec{F}_2, \quad m_3\ddot{\vec{r}}_3 = \vec{F}_3, \tag{5.38}$$

$$\Theta^C\dot{\vec{\omega}} + \vec{\omega} \times \Theta^C \cdot \vec{\omega} = \vec{m}^C. \tag{5.39}$$

The kinematic equation [126]:

$$\begin{bmatrix} \dot{\varphi}_1 \\ \dot{\varphi}_2 \\ \dot{\varphi}_3 \end{bmatrix} = \begin{bmatrix} \dfrac{\cos\varphi_3}{\cos\varphi_2} & -\dfrac{\sin\varphi_3}{\cos\varphi_2} & 0 \\ \sin\varphi_3 & \cos\varphi_3 & 0 \\ -\cos\varphi_3\tan\varphi_2 & \sin\varphi_3\tan\varphi_2 & 1 \end{bmatrix} \begin{bmatrix} \omega_1 \\ \omega_2 \\ \omega_3 \end{bmatrix}. \tag{5.40}$$

The non-zero initial conditions (see Fig. 5.9):

$$y_2 = a + b, \quad y_3 = -a - b, \quad \varphi_3 = \pi/2, \quad \omega_3 = \omega_o,$$

$$\dot{y}_2 = \frac{|\vec{v}_{20}| - |\vec{v}_{30}|}{2}, \quad \dot{y}_3 = -\dot{y}_2. \tag{5.41}$$

If the connections of the rotating system are eliminated and if the braking forces at the exits of the strings from the central body are not acting immediately after separation, the bodies begin to move by inertia as three independent bodies having no effect on each other. Because of the different initial velocities of the bodies the total length of all strings is pulled out. This results in a series of elastic impact and the desired rotation of the deployed system is not reached. However, if the braking forces act on the deploying strings from the central body, the deployed tether starts to rotate. These forces may be realised using Coulomb forces and forces of viscous friction. Let us consider some implementation of them with reference to the system under consideration on the plane of the variables l_i, $\dot{l}_i$. The domain of definition of the magnitude of the braking force in this plane satisfies the conditions $0 \leq l \leq L$, $\dot{l} > 0$, where L is the design length of the string in the operational mode of the systems. Some braking forces F_{fr} which we believe are of practical importance in applications are as follows:

1. Braking force constant in all definitional domain using pure Coulomb friction

$$F_k^{br} = f_0 \quad (k = 1, 2), \tag{5.42}$$

where f_0 is a given value.

2. Braking forces linearly depending on the length of the string during the initial stage of deployment, which is bounded by some intermediate length $\bar{l}$ of the deployed string and is constant on the final

domain of deployment. Only Coulomb friction is used. c_l is a given coefficient.

$$F_k^{br} = \begin{cases} f_0 + c_l\,(\tilde{l} - l) & \forall 0 \le l \le \tilde{l}, \\ f_0 & \forall \tilde{l} \le l \le L, \quad (k = 1, 2). \end{cases} \tag{5.43}$$

3. Braking forces using viscous friction and linearly depending on the velocity $\dot{l}$ of extraction of the string from the container

$$F_k^{br} = c_v\,\dot{l}, \quad (k = 1, 2), \tag{5.44}$$

where c_v is a given coefficient.

4. Braking forces using viscous friction and linear in $\dot{l}$. They do not depend on l at the beginning, but then they are increasing with increasing l

$$F_k^{br} = \begin{cases} c_v\,\dot{l} & \forall 0 \le l \le \tilde{l}, \\ c_v\,\dot{l} + c_{vl}\,\dot{l}\,l & \forall \tilde{l} \le l \le L, \quad (k = 1, 2), \end{cases} \tag{5.45}$$

where c_{vl} is a given coefficient.

5. Braking forces using Coulomb and viscous friction and linear in $\dot{l}$ and constant in l

$$F_k^{br} = c_v\,\dot{l} + f_0, \quad (k = 1, 2). \tag{5.46}$$

6. Braking forces using Coulomb and viscous friction and linear in $\dot{l}$ constant on l in the beginning, then increasing with increasing l

$$F_k^{br} = \begin{cases} c_v\,\dot{l} + f_0 & \forall 0 \le l \le \tilde{l}, \\ c_v\,\dot{l} + c_{vl}\,\dot{l}\,l + f_0 & \forall \tilde{l} \le l \le L, \quad (k = 1, 2). \end{cases} \tag{5.47}$$

Before we begin the description of the results of the numerical simulation of the system deployment, we note that the value of the necessary initial angular velocity of rotation may be determined on the basis of the theorem of moment of momentum in absence of external forces, since the braking forces are internal forces.

5.3.4 Numerical modelling of deployment

On the basis of the constructed mathematical model (5.38)–(5.41) extensive numerical simulation of the system's dynamics of deployment was conducted varying the values of main parameters of the system and the realisation of the braking forces (5.42)–(5.47). We use the following values of the main physical parameters of the system: the masses of bodies: $m_1 = 10\,\text{kg}$, $m_2 = 10\,\text{kg}$, $m_3 = 10\,\text{kg}$; the moments of inertia of bodies with respect to the main axes of inertia: $\Theta_{11} = 1\,\text{kg}\,\text{m}^2$, $\Theta_{22} = 1\,\text{kg}\,\text{m}^2$, $\Theta_{33} = 0$, $L_1 = 10\,\text{m}$, $L_2 = 10\,\text{m}$, tension stiffness of the cable $EF = 0.6106$ N. Numerical simulations were carried out using the fourth-order Runge–Kutta method with an adaptive time step.

The case of pure Coulomb friction, constant in time, gives a mode of deployment possessing characteristic properties. During extraction of the cables from the container there occur significant angular oscillations of the central body similar to rotational oscillations of a rigid body under the action of a restoring moment (curve 1 in Fig. 5.10). Such a moment is created by the braking forces, the directions of which coincide with the directions of the stretched tethers.

While the cable is extracted from the container, this tension force is equal to force of friction.

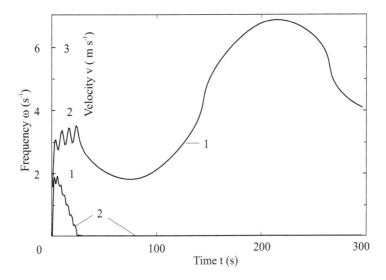

FIGURE 5.10
Angular oscillations of the central body and velocity of string extracting.

At the indicated oscillations, the velocity of pooling out of the tethers also has oscillatory character (curve 2 in Fig. 5.10). When the process of extraction of the tether is completed, the character of the restoring forces essentially changes. They become the forces of elastic tension of the cables arising only if the cables are under tension. The duration of such a condition is extremely small, then the bodies lose the connection again and the tension of the tethers vanishes. Since the momentum of the arising restoring moments in the deployed system is substantially smaller than at the stage of deployment (at the expense of the short-term stretching of the string), the period of the angular oscillations is noticeably greater here. The value of amplitude of angular oscillations is determined by the system condition at the instant of the first stretching the tether by inertial forces after completion of extraction of the cable and can vary within a wide range of limits. Since the initial moment of momentum of the system was normal to the plane Cxy and external forces are absent, the system moves in this plane all the time. This motion is shown in Fig. 5.11.

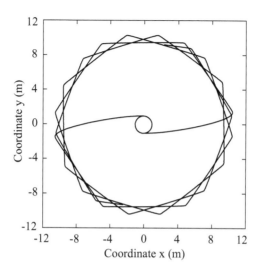

FIGURE 5.11
System behaviour in the plane Oxy of the absolute frame of reference.

The trajectories of the centres of the end bodies contain intervals of time when the strings are stretched. Outside of these intervals, the bodies are moved on inertia and the strings are not stretched.

The formation of the forces on the second version during simulation of deployment of the tether has shown that at the variable braking forces it is possible to achieve more acceptable deployment than in the previous case, but the system behaviour after deployment is still very sensitive to the values of the coefficients in the expressions for the forces. So, at the value c_v equal $0.137\,\mathrm{N\,m^{-1}}$, the tether is deployed quite acceptably on character of further motion, but its length is 4 cm less than nominal length (Fig. 5.12). At c_v equal to $0.135\,\mathrm{N\,m^{-1}}$ (case 1) the tether is deployed on full length, but the velocities of bodies have already nonzero projections on directions of the strings at the instant of termination of the deployment. It results in excitation of angular oscillations of the central body. Decrease of the value c_v up to 0.130 (case 2) results already in other mode of motion (Fig. 5.13), in which the internal resonance in the system between the angular frequency of rotation of the tether and the oscillation frequency of the central body is seen.

Without considering in detail the simulation of the dynamics of deployment at the remaining versions of the friction forces, we note that the results, acceptable in our opinion, are given by the use of only viscous friction, and the relation of the braking force to the length of the deployed string and its

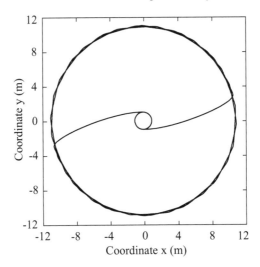

FIGURE 5.12
System behaviour in the plane Oxy of the inertial coordinate frame in the case 1 of constant Coulomb friction.

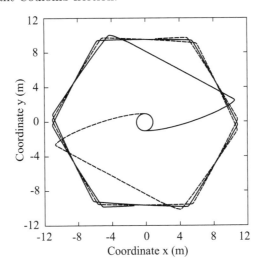

FIGURE 5.13
The system behaviour in the plane Oxy of the inertial coordinate frame in the case 2 of constant Coulomb friction.

feed out velocity may be selected as follows:

$$
F_k^{br} = \begin{cases}
c_{1v}\,\dot{l}(l_1 - l)/l_1 + c_{2v}\dot{l} & \forall 0 \le l \le l_1, \\
c_{2v}\dot{l} & \forall l_1 \le l \le l_2, \\
c_{3v}\,\dot{l}(l - l_2)/(L - l_2) + c_{2v}\dot{l} & \forall l_2 \le l \le L, \quad (k = 1,2)
\end{cases}
$$

c_{1v}, c_{2v}, c_{3v} are given coefficients. Such a structure of braking forces has advantages in comparison with other considered schemes. On the initial period of deployment the forces can be created, which ensure so low velocities of pulling-out of the strings that the bodies of the tether do not exceed the bounds of permissible area. It is especially important when the rotating system is close to the space vehicle. During the second period of deployment when the system moved off the space vehicle at a safe distance, the rather small forces of resistance ensure fast deployment. Finishing period of deployment has the task to decrease to the possible minimum the velocity of pulling-out of the string at the instant $t = T_k$ and to avoid the elastic impacts with the subsequent reflections of the end bodies and disturbance of the configuration of the tether.

The acceptable process of deployment of the tether at such a way of formation of resistance forces is shown in Fig. 5.14. In this case the strings are stretching very soft and some noticeable changes of length of the deployed tether do not happen.

The amplitudes of the elastic oscillations arising at such termination of deployment are shown in Fig. 5.15. As it is visible in the figure, the oscillation amplitude at the total deployed length of the tether does not exceed 1 mm. These oscillations are connected to minor angular oscillations of the central body and, naturally, are accompanied by small changes of the forces of tension of the strings.

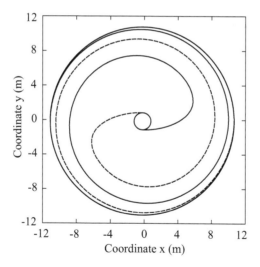

FIGURE 5.14
Process of deployment of the tether in the case of the proposed law of viscous friction.

Thus, consideration of the problem of passive deployment of a tether system of three bodies with the help of the centrifugal forces arising as result of the initial rotation during the pushing away of the tether from the space

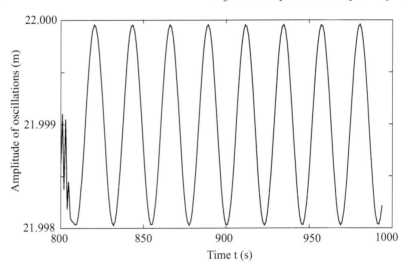

FIGURE 5.15
Amplitudes of elastic oscillations of the end bodies.

vehicle shows that it is possible to achieve deployment of the tether and the angular velocity of the prescribed rotation. It may be done by proper selection of the law of the braking forces in the devices at the exit of the strings.

5.4   Experiment of unreeling the cable

5.4.1   Description of the experiment

The process of unreeling the cable from a reel, which is at rest, is considered by generating the initial velocity of the first body of the small TSS attached to the initial end of the cable. The required length of the unwinding cable, in accordance with the prospective final shape of the TSS, is about 100 m. In Space conditions the process of deployment proceeds practically in conditions of weightlessness and of vacuum.

The purposes of experimental researches of the process of unreeling of the cable are the following: definition of suitability of the design of storing the cable for the creation of small TSS; definition of sufficiently large kinetic energy of motion of the first body for unreeling the cable of given length; calculation of the force of resistance to overcome in unreeling the cable; estimation of the duration of the process of deployment and velocity of the first body at the termination of process of unreeling of the cable.

The process of unreeling of the cable depends on the initial kinematic parameters of motion of the first body, the dynamics of its motion and the dynamics of the cable and certainly on the way the cable is reeled. Features of the way of reeling the cable are expressed dynamically by the force of resistance in unreeling the cable and, probably, by additional disturbances of the dynamics of the cable. The suitability of the way of reeling of a cable for the design of TSS is determined by following: the forces of resistance to unreeling the cable allow to organize the deployment of the system for the prescribed length and there is no danger of entangling (forming spatially complicated configurations) of the cable.

If we do not take into account the dynamics of the already unreeled part of the cable, i.e., to consider the part of the cable located between the bodies as massless, the equation of motion of the mass centre of the first body along a straight line looks like

$$m_1 \ddot{x} = -T_c, \tag{5.48}$$

where m_1 is the mass of the first body, x is the coordinate of its centre of mass along a straight line of motion, T_c is the value of the force of resistance against unreeling of the cable.

Let us consider now the fly-wheel (Fig. 5.16) unreeling the same cable as the body. The equations of its motion look like

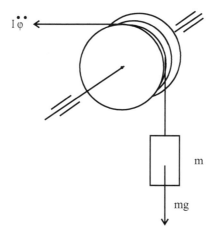

FIGURE 5.16
Scheme of experiment on determination of moment of inertia of fly-wheel.

$$J \ddot{\varphi} = -r_k T_c, \tag{5.49}$$

where J is the moment of inertia, φ is angle around of rotation axis, r_k is the radius of the reel of the fly-wheel where the cable is stored.

It is easy to see that for initial conditions $r \dot{\varphi}_0 = \dot{x}_0$ and for $J/r_k^2 = m_1$ the models (5.48) and (5.49) are dynamically equivalent. Hence the velocities of unreeling of a cable in both cases is identical for all time.

The creation of the device to realise the model (5.49) is much easier to do than the realisation of model (5.48). It is because the aerodynamic resistance to rotation of a symmetrical rigid body about the axis of symmetry for fixed mass centre practically is equal to zero. And also because to get small friction in bearings is much easier to model than to model weightlessness of the moving body with length of 100 m.

The use of the fly-wheel for experimental research of the process of unreeling of the cable allows to achieve the purposes of the research. Really, the fly-wheel allows to make full unreeling of the cable with the velocity close to the prescribed one and with the prescribed inertial forces (because $J/r_k^2 = m_1$). It allows to solve the problem about suitability of the way of reeling of the cable and to determine the sufficiently large kinetic energy of motion of the first body for the full unreeling of the cable. Calculation of friction in bearings allows to calculate the force of resistance against unreeling of the cable for various velocities of unreeling and, hence, to obtain time estimations of the process of deployment and the velocity of motion of the first body at termination of unreeling of the cable.

The scheme of realisation of measurements in the experiment may be the following. On the fly-wheel on which the cable is reeled up, marks are placed on the edge of the fly-wheel of the angular distance of π rad. On the fixed part there is a non-contact sensor, such as a travelling switch, which changes its state at the moment of passage of the mark near it. Thus, in the experiment the process of the test is shown as a sequence of events in time, corresponding to the rotation of the fly-wheel by an angle of π rad. Practically the measurement equipment does not have an influence on the experimental process.

To perform the experiment it is necessary only to fix the time of change of the state of the switch. However, it is not necessary to fix the number of the events, because of the pre-determined sequences of their changes.

One of the features of the studied process of unreeling of the cable is the necessity of registration of time of the events with an accuracy not below 0.001 seconds, if the frequency of the events is about 10 to 50 hertz during a period of 50 to 100 seconds.

So, the monitoring system must be developed including the non-contact sensor, the controller and the computer. Impulses of sensors are registered by the controller supplied with both autonomous and external memory and the driver for the interaction with the computers.

To perform the experiments of unreeling of the cable we must evaluate the moments of inertia of the fly-wheel, the forces of friction in bearings and the forces of resistance against unreeling of the cable.

According to the program of research of the process of unwinding [113], the motion of the cable due to one separating body of the TSS is simulated in the ground experimental research by the rotation of a fly-wheel. Therefore the definition of the forces of resistance against unwinding of the cable requires the determination of the moment of inertia of the fly-wheel and the forces of friction in the bearings. All techniques cited below were developed for the specific equipment and on the basis of specific experimental data. On the basis of the qualitative analysis of the experimental data the general assumptions about the character of the forces acting on the process were performed and then by the way of minimization of quadrates of deviations of the calculated values from the experimental data the parameters of these force actions were determined.

5.4.2 Moment of inertia of a fly-wheel

The moment of inertia of a fly-wheel was determined on the basis of the experiment, the model of which is presented in Fig. 5.16.

The equations of motion look like

$$\begin{aligned} J\ddot{\varphi} &= rT - M, \\ m\ddot{x} &= mg - T. \end{aligned} \tag{5.50}$$

Here T is the tension force in the cable, M is the moment of the forces of friction in the bearings.

Provided the cable is sufficiently stiff, it is possible to assume that $\dot{x} = r\dot{\varphi}$. Hence

$$\left(J + mr^2\right)\ddot{\varphi} = rmg - M. \tag{5.51}$$

As results of the experiments show, at comparatively small velocities of rotation, the moment of friction forces in the bearings can be neglected. Then the variation of the angle φ is given by the formula

$$\varphi = \frac{rmg}{J + mr^2}\frac{t^2}{2} + \dot{\varphi}_0 t + \varphi_0, \tag{5.52}$$

and by virtue of arrangement of the experiment it is possible to suppose that the initial angular velocity is equal to zero ($\dot{\varphi}_0 = 0$). Since the conditions of realisation of the experiment do not allow to determine precisely the time of beginning of the motion (or the initial value of the angle φ_0), the difference between the increment of the angle by formula (5.52) and its real increment was considered as a function to be minimized. The real increment has the form

$$F = \sum_{i=1}^{N} \left[(\varphi(t_{i+1}) - \varphi(t_i)) - (\varphi_{i+1} - \varphi_i)\right]^2,$$

where $\varphi(t_i)$ is the value of the angle of rotation calculated by the formula (5.52) in the experimentally determined instants of time, φ_i is the real value of the angle of rotation at the time instants, N is the number of performed experiments.

As a result of minimizing the function F with respect to the moment of inertia J we obtain

$$\frac{\partial F}{\partial J} = 0,$$

$$\sum_{i=1}^{N} \left[\frac{rmg}{J + mr^2} \frac{1}{2} \left(t_{i+1}^2 - t_i^2 \right) - 2\pi \right] \left(t_{i+1}^2 - t_i^2 \right).$$

Hence,

$$J^* = \frac{rmg \sum_{j=1}^{N} \left(t_{i+1}^2 - t_i^2 \right)^2}{4\pi \sum_{i=1}^{N} \left(t_{i+1}^2 - t_i^2 \right)} - mr^2, \tag{5.53}$$

where it is assumed that $\varphi_{i+1} - \varphi_i = 2\pi$.

In Fig. 5.17 the graphs of variation of angle φ are presented. The triangles show values of the angle φ obtained from the experiment, the lines are the graphs constructed from formulas (5.52), (5.53).

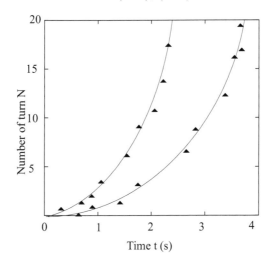

FIGURE 5.17
Experimental and computational values of angle ϕ at determination of hand-wheel moment of inertia.

5.4.3 Moment of friction forces in the bearings

The definition of the moment of the forces of resistance against rotation of a fly-wheel was carried out on the basis of the experimental data of free (inertial) rotation of the fly-wheel. As unknown parameters the coefficients of the constant moment of friction and the moment of friction proportional to the velocity of rotation were considered.

Hence the resulting equation of rotation of the fly-wheel has the form

$$\ddot{\varphi} = -M_0 - c\dot{\varphi} \tag{5.54}$$

and its solution at $\varphi_0 = 0$ can be written as

$$\varphi = \frac{(M_0/c + \dot{\varphi}_0)}{c} \left(1 - e^{-ct} \right) - \frac{M_0}{c} t. \tag{5.55}$$

Since friction is small, the definition of the initial angular velocity can be carried out with high accuracy:

$$\dot{\varphi}_0 = 2\pi/T,$$

where T is time of one complete rotation of the fly-wheel.

The values of the unknown parameters c and M_0 were determined by way of minimization of a sum of quadrates of deviations of the solution (5.50) from the experimental values:

$$F_1 = \sum_{i=1} N(\varphi(t_i) - \varphi_i)^2 \quad \rightarrow \quad \left\{ \begin{matrix} MIN \\ c, M_0 \end{matrix} \right\}. \tag{5.56}$$

The minimization of the function was carried out numerically. As a result the following values were obtained: $M_0 = 0.006\,\text{s}^{-2}$, $c = 0.026\,\text{s}^{-1}$. Results of verifying calculations are presented in Fig. 5.18. The triangle points represent data of experiments, and the solid lines correspond to the formula (5.55).

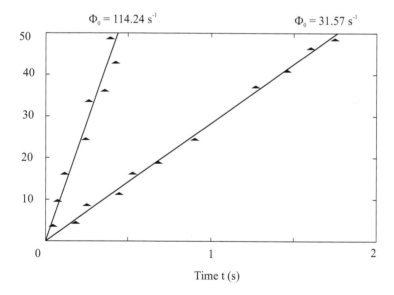

FIGURE 5.18
Results of test computations.

5.4.4 Resistance force against unreeling of cable

As it was remarked above, the evaluation of the forces of resistance against unwinding of the cable from a reel should allow to determine the energy needed for the complete unwinding of the cable, the time of unwinding and the final velocity of the body at the end of this process. The definition of these parameters is necessary for the improvement of the requirements for the device of pushing off and for the device to brake to zero the velocity of the body.

The definition of the physical nature of the forces of resistance against unwinding and the influence of other parameters on the value of this resistance requires a special investigation. Since the mass of the cable can have the value of the order of 1% of the mass of the body, it is necessary to consider the motion of the body in the process of unwinding as motion of a body with variable mass. In addition, it is necessary to determine the dependence of the forces of friction on the velocity of the body (whether it is linear or non-linear and determine the character of the non-linearity), on the mass of the cable, on the size of the reel and on the methods of winding, etc [41].

The presented investigation is limited by the assumption that the force of resistance against unwinding consists of a force which consists of a constant part and a part proportional to the velocity of unwinding. In this case the motion of the fly-wheel is described by the equation of the form (5.51), in which the constants determine the total value of the moment created by the force of resistance against unwinding of the cable and the force of friction of the fly-wheel. It is necessary to check the validity of this assumption by the way of comparison of results of calculations and experiments.

By conducting the *a priori* definition of the parameters M_0 and c for the fly-wheel without unwinding of the cable from the reel, and then by repeating the same procedure during unwinding of the cable, it is possible to obtain values of the parameters ΔM_0 and Δc describing only the moment of the force of resistance against unwinding. By virtue of the assumption about a linear dependence of the force of resistance on the velocity of unwinding we obtain the values ΔM_0 and Δc as the difference of values M_0 and c in two experiments.

Then the motion of the fly-wheel under the action only of the moment from the force of resistance against unwinding looks like

$$\ddot{\varphi} = -\Delta M_0 - \Delta c \dot{\varphi}. \tag{5.57}$$

The motion of the body with mass $m = J/r^2$ under action of the force of resistance against unwinding is described by the equation

$$m\ddot{x} = -\frac{J}{r}\Delta M_0 - \frac{J}{r^2}\Delta c \dot{x}. \tag{5.58}$$

The expression on the right-hand side represents the force of resistance against unwinding of the cable, which can be used for the calculation of the motion of the body after it is pushed off.

Bibliography

[1] N. I. Alekseev. *Statics and steady motion of a flexible string [In Russian]*. Legkaja industrija, Moscow, 1970.

[2] A. A. Alifov and K. V. Frolov. *Interaction of non-linear vibratory systems with a power source [In Russian]*. Nauka, Moscow, 1985.

[3] A. Alpatov, V. Dranovskii, V. Khoroshilov, A. Pirozhenko, and A. Zakrzhevskii. Research of dynamics of space cable systems stabilised by rotation, iaf-97-a.2.10. page 13, Turin, Italy, October 1997. IAF.

[4] A. P. Alpatov, V. V. Beletsky, V. I. Dranovskii, A. E. Zakrzhevskii, A. V. Pirozhenko, and V. S. Khoroshilov. Dynamics of small space tether systems stabilised by rotation [In Russian]. *Tekhnicheskaja mekhanika*, (1):85–100, 2001.

[5] A. P. Alpatov, P. A. Belonozhko, A. V. Pirozhenko, and V. A. Shabokhin. About evolution of rotational motion of a tether of two bodies on orbit [In Russian]. *Kosmicheskie issledovanija*, 28(5):692–701, 1990.

[6] A. P. Alpatov, V. I. Dranovskii, A. E. Zakrzhevskii, A. V. Pirozhenko, and V. S. Khoroshylov. Space tethered systems. The review of the problem [In Russian]. *Kosmicheskaja nauka i tekhnologija*, 3:12–22, 1997.

[7] A. P. Alpatov and A. E. Zakrzhevskii. Deployment of an elastic tether in result of initial twist. pages 21–22, Russia, St Petersburg (Repino), June 27–July 6 2002. Inst. for Problems in Mech. Engng, Russ. Acad. Sci.

[8] A. P. Alpatov, A. E. Zakrzhevskii, and G. Matarazzo. Deployment of a three body tether in field of centrifugal forces [In Russian]. *Tekhnicheskaja mekhanika*, (2):3–12, 2002.

[9] Ja. L. Alpert. *Waves and artificial bodies in near-earth plasma [In Russian]*. Nauka, Moscow, 1974.

[10] Ja. L. Alpert, A. V. Gurevich, and L. P. Pitaevskij. *Artificial satellites in rarefied plasma [In Russian]*. Nauka, Moscow, 1964.

[11] F. Angrilli, G. Bianchini, M. Da Lio, and G. Fanti. Modeling the mechanical properties and dynamics of the tethers for the TSS-1 and TSS-2 missions. *ESA J.*, (12):354–368, 1988.

[12] F. Angrilli, P. Forno, and F. Reccannello. Effects of several tethers in space. *J. Spacecraft and Rockets*, 34(2):239–245, 1997.

[13] F. Angrilli, R. Da Forno, and B. Saggin. A new full non-linear model of the tethered satellite system based on the characteristics method. pages 1171–1180, Washington, D.C., 1995.

[14] P. Appell. *Traite de mecanique rationnelle*, volume 1 and 2. Paris, 6th edition, 1953.

[15] V. I. Arnold. *Mathematical methods of classical mechanics [In Russian]*. Nauka, Moscow, 1983.

[16] V. V. Beletsky. *Motion of an artificial satellite about a centre of mass*. Israel Programm for Scientific Translations, Jerusalem, 1966.

[17] V. V. Beletsky. About relative motion of a tether of two bodies on an orbit. ii [In Russian]. *Kosmicheskie issledovanija*, 7(6):827–840, 1969.

[18] V. V. Beletsky. *Motion of a satellite about a centre of mass in a gravitational field [In Russian]*. Nauka, Moscow, 1975.

[19] V. V. Beletsky. *Applied problems of dynamic billiard [In Russian]*, volume 19, pages 204–239. Fizmatlit, Moscow, 2001. special issue.

[20] V. V. Beletsky. *Essays on the Motion of Celestial Bodies*. Birkhauser, Basel, 2001.

[21] V. V. Beletsky, V. I. Dranovskii, V. S. Khoroshylov, Ju. D. Saltykov, A. P. Alpatov, A. V. Pirozhenko, and A. E. Zakrzhevskii. Chaos regimes and synchronization of motions in dynamics of space tethered systems. iaf-98-a.5.02. page 8, Melbourne, Australia, September 28–October 2 1998. IAF.

[22] V. V. Beletsky and A. V. Grushevskii. The evolution of the rotational motion of a satellite under the action of a dissipative aerodynamic moment. *J. App. Math. and Mech.*, 58(1):11–19, 1994.

[23] V. V. Beletsky and A. M. Janshin. *Influence of aerodynamic forces on rotation of artificial satellites [In Russian]*. Naukova dumka, Kiev, 1984.

[24] V. V. Beletsky and G. V. Kasatkin. Tether of two bodies in orbit. stable of one-link periodic trajectories [In Russian]. *Izvestija of Tula State University*, 2(2):17–22, 1996.

[25] V. V. Beletsky and E. M. Levin. *Dynamics of space tether systems*. American Astronautical Society, San Diego, 1993.

[26] V. V. Beletsky and E. T. Novikova. About relative motion of a tether of two bodies on orbit [In Russian]. *Kosmicheskie issledovanija*, 7(3):377–384, 1969.

[27] V. V. Beletsky and D. V. Pankova. Tether of two bodies on orbit as dynamic billiard [In Russian]. Preprint of Keldysh Inst. of appl. math, 1995.

[28] V. V. Beletsky and D. V. Pankova. Connected bodies in the orbit as dynamic billiard. *Regular and chaotic dynamics*, 1(2):47–58, 1996.

[29] V. V. Beletsky and D. V. Pankova. Effect of aerodynamics on relative motion of orbital tether of two bodies. Part 2. Chaotic and regular motions [In Russian]. Preprint of M. V. Keldysh Institute of applied mathematics RAS, No.40, 1996.

[30] V. V. Beletsky, V. L. Vorontsova, L. M. Kof, and D. V. Pankova. Effect of aerodynamics on relative motion of orbital tether of two bodies. Part 1. Regular motions [In Russian]. Preprint of M. V. Keldysh Institute of applied mathematics RAS, No.38, 1996.

[31] S. Bergamashi, P. Zanetti, and C. Zottarel. Non-linear vibrations in the tethered satellite system-mission. *J. Guidance, Control, and Dynamics*, 19(2):289–296, 1996.

[32] N. N. Bogoljubov and Y. A. Mitropolsky. *Asymptotic methods in the theory of non-linear oscillations*. Hindustan Publishing, Delhi, 1961.

[33] V. V. Bolotin, editor. *Vibrations in engineering. Oscillations of machines, designs and their units [In Russian]*, volume 1. Mashinostroenie, Moscow, 1978.

[34] J. V. Breakwell and J. W. Gearhant. Pumping a tethered configuration to boost its orbit around an oblate planet. *J. Astronaut. Sci.*, 35(1):19–40, 1987.

[35] C. V. L. Charlier. *Die Mechanik des Himmels. Vorlesungen*. Walter de Gruyter, Berlin, 1927.

[36] F. L. Chernous'ko. About motion of a rigid body with mobile masses. *Mechanics of Solids*, (4):33–44, 1973.

[37] F. L. Chernous'ko. On the motion of a solid body with elastic and dissipative elements. *J. App. Math. and Mech.*, 41(1):32–41, 1978.

[38] G. Colombo, S. Bergamaschi, and F. Bevilacqua. The italian participation to the tethered satellite system. *Acta Astronautica*, 8(6-7):353–358, 1982.

[39] M. L. Cosmo, E. C. Lorenzini, and G. E. Gullahorn. Acceleration levels and dynamic noise on seds end-mass. pages 747–760, Washington, D.C.

[40] M. L. Cosmo and E.C. Lorenzini. *Tethers in Space Handbook.* Smithsonian Astrophysical Observatory, 3rd edition, 1997.

[41] E. B. Crellin, F. Janssens, D. Poelaert, W. Steiner, and H. Troger. On balance and variational formulations of the equations of motion of a body deploying along a cable. *J. Appl. Mech.*, 64(2):369–374, 1997.

[42] S. A. Crist and J. G. Eisley. Cable motion of a spinning spring-mass system. *J. Spacecraft and Rockets*, 7(11):1352–1357, 1970.

[43] L. V. Dokuchaev and G. G. Efimenko. Influence of atmosphere to relative motion of a tether of two bodies on orbit [In Russian]. *Kosmicheskie issledovanija*, 10(1):57–65, 1972.

[44] V. Dranovskii, A. Alpatov, V. Khoroshilov, A. Pirozhenko, and A. Zakrzhevskii. Research of dynamics of space cable systems stabilised by rotation. iaf-97-a.2.10. page 13, Turin, Italy, October 1997. IAF.

[45] G. N. Duboshin. *Celestial mechanics. Primal problems and methods [In Russian].* Nauka, Moscow, 1975.

[46] G. N. Duboshin, editor. *The reference book on celestial mechanics and astrodynamics [In Russian].* Nauka, Moscow, 1976.

[47] B. E. Gilchrist, R. Heelis, W. J. Raitt, C. Rupp, H. G. James, C. Bonifazi, K.-I. Oyama, G. Wood, L. M. Brace, and G.R. Carignan. Space tethers for ionospheric-thermospheric-mesospheric science. pages 221–226, Washington, D.C., 1995.

[48] V. P. Glushko, editor. *Astronautics [In Russian].* Sovetskaja entsiklopedija, Moscow, 1985.

[49] O. A. Goroshko and G. N. Savin. *Introduction to mechanics of deformable one-dimensional bodies of a variable length [In Russian].* Naukova Dumka, Kiev, 1971.

[50] X. Gou, X. Ma, and C. Shao. Dynamical characteristics analysis of tethered subsatellite in the presence of offset. *J. Spacecraft and Rockets*, 33(6-8):829–835, 1996.

[51] E. A. Grebennikov. *Method of averaging in the applied problems [In Russian].* Nauka, Moscow, 1986.

[52] V. I. Guljaev, P. P. Lizunov, and N. N. Prudenko. Non-linear oscillations of system of two bodies about a centre of mass on elliptical orbit [In Russian]. *Kosmicheskie issledovanija*, 22(2):165–170, 1984.

[53] K. V. Holshevnikov and V. E. Markov. About stabilisation of percenter altitude [In Russian]. *Kosmicheskie issledovanija*, 13(2):181–184, 1975.

[54] A. P. Ivanov. Collision-free motion in systems with non-retaining constraints. *J. Appl. Math. and Mech.*, 56(1):1–12, 1992.

[55] A. P. Ivanov and A. B. Bazijan. Studying of asymptotic motions of two bodies tether on circular orbit [In Russian]. *Kosmicheskie issledovanija*, 33(5):485–490, 1995.

[56] V. A. Ivanov and Yu. S. Sitarsky. *Flight dynamics of systems of elastically connected space bodies [In Russian]*. Mashinostroenie, Moscow, 1986.

[57] H. G. James, A. W. Yau, and G. Tyc. Space research in the biceps experiment. pages 1585–1598, Washington, D.C., 1995.

[58] T. R. Kane and P. A. Levinson. Deployment of a cable-supported payload from an orbiting spacecraft. *J. Spacecraft and Rockets*, 14(7):409–413, 1977.

[59] D.A. Kiroudis and A. B. Konvej. Advantages of injection of a satellite from elliptical orbit through a cable [In Russian]. *Aerokosmicheskaja tehnika*, (4):198, 1989.

[60] V. I. Komarov. Libratory oscillations of a heavy string in a central field [In Russian]. *Kosmicheskie issledovanija*, 10(1):46–56, 1972.

[61] O. V. Kozlov. *The electrical probe in plasma [In Russian]*. Atomizdat, Moscow, 1969.

[62] L. D. Landau and E. M. Lifshits. *The Classical Theory of Fields*. Butterworth-Heinemann, 4th edition, 1980.

[63] D. D. Lang and R. K. Nolting. Operations with tethered space vehicles. pages 55–66, Houston, Texas, February 1967. NASA. SP-138.

[64] B. J. Lazan. *Damping of materials and members in structural mechanics*. Pergamon Press, 1968.

[65] V. I. Levantovsky. *Mechanics of space flight in an elementary presentation [In Russian]*. Nauka, Moscow, 1980.

[66] L. Liangdon and P. M. Bainum. Effect of tether flexibility on tethered shuttle subsatellite stability and control. *J. Guidance, Control, and Dynamics*, 12(6):874–879, 1989.

[67] A. Lihtenberg and M. Liberman. *Regular and stochastic dynamics*. Springer-Verlag, New York, 2nd edition, 1992.

[68] M. Z. Litvin-Sedoj. *Mechanics of systems of connected rigid bodies. The totals of science and engineering. Series "General mechanics" [In Russian]*. VINITI, Moscow, 1982.

[69] A. Luongo and F. Vestroni. Periodic free oscillations of a tethered satellite systems. *J. Sound and Vibration*, 175(2):299–315, 1994.

[70] A. I. Lurie. *Analytical Mechanics*. Springer, 2002.

[71] A. P. Markeev. To dynamics of elastic body in a gravitational field [In Russian]. *Kosmicheskie issledovanija*, 27(2):163–175, 1989.

[72] A. P. Markeev. *Changes of aircraft attitude of fast rotations of viscoelastic cylindrical shell in gravitational field [In Russian]*. AN SSSR, Moscow, 1990.

[73] V. E. Markov. Parrying of external disturbances on a space vehicle by a method of change of its geometry of masses. *Mechanics of Solids*, (5):3–9, 1974.

[74] M. Martinez-Sanchez and S. E. Gevit. Orbital modification using forced tether-length variations. *J. Guidance, Control, and Dynamics*, 10(3):233–241, 1987.

[75] E. P. Mazets. *Micrometeorites in space. Dust in atmosphere and near-earth space [In Russian]*. Nauka, Moscow, 1973.

[76] D. R. Merkin. *Introduction to a mechanics of a flexible string [In Russian]*. Nauka, Moscow, 1980.

[77] R. E. Mickens and M. Mixon. Application of generalized harmonic balance to an anti-symmetric quadratic non-linear oscillator. *J. Sound and Vibration*, 159(3):546–548, 1992.

[78] A. K. Misra and V. J. Modi. Three-dimensional dynamics and control of tethered connected n-body systems. *Acta Astronautica*, 26(2):77–84, 1992.

[79] A. K. Misra, V. J. Modi, G. Tyc, F. Vegneron, and A. Jablonski. Dynamics of low-tension spinning tethers. pages 1309–1324, Washington, D.C., 1995.

[80] Yu. A. Mitropol'sky. *Method of averaging in non-linear mechanic [In Russian]s*. Naukova Dumka, Kiev, 1971.

[81] Ju. G. Mizun. *Ionosphere of the Earth [In Russian]*. Nauka, Moscow, 1985.

[82] N. N. Moiseev. *Asymptotic methods of non-linear mechanics [In Russian]*. Nauka, Moscow, 1969.

[83] Nam Tum Po. Influence of aerodynamic braking to motion of a spherical satellite about a centre of mass. *Bull. of Institute of theoretical astronomy*, 10(5):84–91, 1965.

[84] L. G. Napolitano and F. Bevilacqua. Tethered constellations, their utilization as microgravity platforms and relevant features. Lausanne, Switzerland, October 1984. IAF. IAF-84-439.

[85] L. Pars. *A Treatise on Analytical Dynamics*. Ox Bow Press, 1981.

[86] B. Paul. Planar librations of an extensible dumbbell satellite. *AIAA J.*, 1(2):411–418, 1963.

[87] A. V. Pirozhenko. Spatial motion of two bodies with an elastic non-restraining coupling. *Int. Appl. Mech.*, 25(11):1153–1159, 1989.

[88] A. V. Pirozhenko. Equations of disturbed motion of a mass point on elastic connection [In Russian]. *Prikladnaja mekhanika*, 26(5):126–129, 1990.

[89] A. V. Pirozhenko. Mission control of a tether of two bodies in a newtonian field of forces by change of length of connection. *Kosmicheskie issledovanija*, 30(4):50–56, 1990.

[90] A. V. Pirozhenko. The calculation of a first approximation for systems with strongly non-linear oscillatory sections. *J. Appl. Math. and Mech.*, 57(2):267–273, 1993.

[91] A. V. Pirozhenko. Motion control of a tether of two bodies about the mass centre [In Russian]. *Tekhnicheskaja mekhanika*, 26(1):31–37, 1993.

[92] A. V. Pirozhenko. About influence of dissipation of energy in a material of a string on evolution of rotational motion of a space tether system [In Russian]. *Kosmichna nauka i tehnologija*, 4(5-6):2–9, 1998.

[93] A. V. Pirozhenko. Derivation of the new forms of perturbed keplerian motion [In Russian]. *Kosmichna nauka i tehnologija*, 5(2-3):103–107, 1999.

[94] A. V. Pirozhenko. Chaotic regimes of motion in the dynamics of space tethered systems. 1. Analysis of the problem [In Russian]. *Kosmichna nauka i tehnologija*, 7(2-3):83–89, 2001.

[95] A. V. Pirozhenko. Chaotic regimes of motion in the dynamics of space tethered systems. 2. Mechanical image of the phenomenon [In Russian]. *Kosmichna nauka i tehnologija*, 7(2-3):90–99, 2001.

[96] A. V. Pirozhenko. Chaotic regimes of motion in the dynamics of space tethered systems. 3. Effect of the energy dissipation [In Russian]. *Kosmichna nauka i tehnologija*, 7(5-6):13–20, 2001.

[97] G. S. Pisarenko, A. P. Jakovlev, and V. V. Matveev. *Vibro-absorbing properties of design materials [In Russian]*. Naukova Dumka, Kiev, 1971.

[98] R. T. Pringle. On the exploitation of non-linear resonance in damping an elastic dumbbell satellite. *AIAA J.*, 6(7):1217–1222, 1968.

[99] C. C. Rupp. A tether tension control law for tethered subsatellites deployed along local vertical. Technical Report NASA TMX–64963, Marshall Space Flight Center, September 1975.

[100] N. V. Saltykov. *Flexible strings in flows [In Russian]*. Naukova dumka, Kiev, 1974.

[101] N. A. Savich. Central problem of space electronics and electrodynamic-TSS. pages 909–920, Washington, D.C., 1995.

[102] M. Schagerl, A. Steindl, and H. Troger. Dynamical analysis of the deployment process of tethered satellite systems. pages 345–354. IUTAM, Kluwer Academic Publishers, 2000.

[103] S. T. Sergeev. *Steel cables [In Russian]*. Tehnika, Kiev, 1974.

[104] V. S. Shchedrov. *Fundamentals of a mechanics of a flexible string [In Russian]*. Mashgiz, Moscow, 1961.

[105] E. N. Sinitsin. Evolution of keplerian motion of a visco-elastic planet [In Russian]. *Astronomicheskij zhurnal*, 60(3):630–635, 1990.

[106] E. N. Sinitsin. About influence of visco-elastic properties of a material of a body to its fast rotations in a gravitational field. *Mechanics of Solids*, (1):31–38, 1993.

[107] *Standard atmosphere. GOST 4401-81 [In Russian]*. Standarty, Moscow, 1981.

[108] W. Steiner, A. Steindl, and H. Troger. Dynamics of a space satellite system with two rigid endbodies. pages 1367–1379, Washington, D.C., 1995.

[109] W. Steiner, J. Zemann, A. Steindl, and H. Troger. Numerical study of large amplitude oscillations of two satellites continuous tether system with varying length. iaf-93-a.3.22. 1993.

[110] W. Steiner, J. V. Zemann, A. Steindl, and H. Troger. Numerical study of large amplitude oscillations of a two satellites continuous tether system with varying length. *Acta Astronautica*, 35:607–621, 1995.

[111] T. G. Strizhak. *Methods of research of dynamic systems of type pendulum [In Russian]*. Nauka, Moscow, 1981.

[112] F. Tisserand. *Trait de mecanique celeste*, volume 1. Paris, 1889.

[113] H. Troger, A. P. Alpatov, V. V. Beletsky, V. I. Dranovskii, A. V. Pirozhenko, V. S. Khoroshilov, and A. E. Zakrzhevskii. Experimental and computational analysis of tethered space systems. Phase 1/ setting of the problems for research and preparation of the input data. Technical Report 94-0644, INTAS, 1997.

[114] G. Tyc, C. C. Rupp, A. M. Jablonski, and F. R. Vigneron. A shuttle deployed tether technology demonstration mission to serve canadian and united states needs. pages 1599–1610, Washington, D.C., 1995.

[115] Yu.G. Markov V. G. Vil'ke. Evolution of translational-rotational motion of visco-elastic planet in a central field of forces [In Russian]. *Astronomicheskij Zhurnal*, 65(4):861–867, 1988.

[116] B. I. Valjaev and G. L. Kozhevnikova. *Cable systems in a fluid flow. The review [In Russian]*, volume 489. TsAGI, Moscow, 1976.

[117] J. C. van der Ha. Orbital and relative motion of a tethered satellite system. *Acta astronautica*, 12(4):207–211, 1985.

[118] V. G. Vil'ke. Motion of a visco-elastic sphere in a central Newtonian force field [In Russian]. *Kosmicheskie issledovanija*, 17(3):280–284, 1979.

[119] V. G. Vil'ke. Motion of visco-elastic full-sphere in a central Newtonian field of forces. *J. Appl. Math. and Mech.*, 44(3):395–402, 1980.

[120] V. G. Vil'ke and K. M. Lebedev. Resonant phenomena at evolution of translational-rotational motion of visco-elastic planet [In Russian]. *Kosmicheskie issledovanija*, 25(1):148–152, 1987.

[121] O. L. Voloshenjuk and A. V. Pirozhenko. Technique of research of perturbed motion of space systems [In Russian]. *Tekhnicheskaja mekhanika*, (1):18–27, 2000.

[122] O. L. Voloshenjuk and A. V. Pirozhenko. To calculation of dissipation of essentially non-linear longitudinal oscillations space tethered system stabilised by rotation [In Russian]. *Tekhnicheskaja mekhanika*, (2):3–12, 2000.

[123] V. M. Volosov and B. I. Morgunov. *Method of averaging in theory of non-linear vibratory systems [In Russian]*. MGU, Moscow, 1971.

[124] G. Wiedermann, M. Schagerl, A. Steindl, and H. Troger. Computation of force controlled deployment and retrieval of a tethered satellite system by the finite element method. page 20, 1999.

[125] E. T. Wittaker. *Treatise on the Analytical Dynamics of Particles and Rigid Bodies*. Cambridge Univ. Press, 1927.

[126] J. Wittenburg. *Dynamics of Systems of Rigid Bodies.* B. G. Teubner, Stuttgart, 1977.

[127] A. E. Zakrzhevskii. Optimal control for elastic space construction. *ZAMM*, 78, S3:1135–1136, 1998.

[128] A. E. Zakrzhevskii. The dynamics of systems of rigid and elastic bodies as applied to spacecraft. *Int. Appl. Mech.*, 36(8):1573–1582, 2000.

[129] A. E. Zakrzhevskii. Slewing of flexible spacecrafts with minimal relative flexible acceleration. *J. Guidance, Control, and Dynamics*, 31(3):563–570, 2008.

[130] G. M. Zaslavskij and R. Z. Sagdeev. *Introduction to non-linear physics: From a pendulum up to turbulence and chaos [In Russian].* Nauka, Moscow, 1986.

[131] A. I. Zhirjakov. *Research of oscillation damping and stiffness of steel cables [In Russian].* PhD thesis, The Kharkov highest engineering school the name of N. I. Krylov, Kharkov, 1972.

Index

Printed and bound by CPI Group (UK) Ltd, Croydon, CR0 4YY

29/10/2024

01780531-0001